CONTENTS

Foreword by William J. Crowe — vii

Introduction: The Promise of Immaculate Warfare — 1
Stephen D. Wrage

1. Air Power and the Coercive Use of Force — 5
Scott A. Cooper

2. Air Power Strategy and the Problem of Coercion — 21
Spencer Abbot

3. Coalition Warfare: The Commander's Role — 51
Derek S. Reveron

4. The Politics of Air Strikes — 71
Scott A. Cooper

5. The Ethics of Precision Air Power — 85
Stephen D. Wrage

6. Conclusion — 101
Stephen D. Wrage

Further Reading — 113

Index — 115

About the Contributors — 119

IMMACULATE WARFARE

Participants Reflect on the Air Campaigns over Kosovo, Afghanistan, and Iraq

Edited by Stephen D. Wrage

Westport, Connecticut
London

Library of Congress Cataloging-in-Publication Data

Immaculate warfare : participants reflect on the air campaigns over Kosovo, Afghanistan, and Iraq / edited by Stephen D. Wrage ; foreword by William J. Crowe.
    p. cm.
Includes bibliographical references and index.
ISBN 0-275-97643-2 (alk. paper) — ISBN 0-275-97644-0 (pbk. : alk paper)
1. Precision bombing—United States. 2. Guided bombs. 3. Guided missiles—United States. 4. Kosovo (Serbia)—History—Civil War, 1998—Aerial operations, American. 5. Afghanistan—History—2001- 6. Air power—United States. I. Wrage, Stephen D.
UG703.I46 2003
358.4'2'0973—dc21   2003045768

British Library Cataloguing in Publication Data is available.

Copyright © 2003 by Stephen D. Wrage

All rights reserved. No portion of this book may be
reproduced, by any process or technique, without the
express written consent of the publisher.

Library of Congress Catalog Card Number: 2003045768
ISBN: 0-275-97643-2

First published in 2003

Praeger Publishers, 88 Post Road West, Westport, CT 06881
An imprint of Greenwood Publishing Group, Inc.
www.praeger.com

Printed in the United States of America

The paper used in this book complies with the
Permanent Paper Standard issued by the National
Information Standards Organization (Z39.48-1984).

Copyright Acknowledgments

The editor and publisher gratefully acknowledge permission for use of the following material:

Chapter 1 is reprinted from Scott A. Cooper, "Air Power and the Coercive Use of Force." *The Washington Quarterly* 24(4) (Autumn, 2001), pp. 81-93. © 2001 by the Center for Strategic and International Studies (CSIS) and the Massachusetts Institute of Technology.

Chapter 3 is reprinted from Derek Reveron, "Coalition Warfare: The Commander's Role." *Defense & Security Analysis* 18(2) (2002): 107-121. http://www.tandf.co.uk.

Chapter 4 is reprinted from Scott Cooper, "Politics of Air Strikes." *Policy Review*. (June-July 2001) (Issue 107), pp. 55-66.

P

In order to keep this title in print and available to the academic community, this edition was produced using digital reprint technology in a relatively short print run. This would not have been attainable using traditional methods. Although the cover has been changed from its original appearance, the text remains the same and all materials and methods used still conform to the highest book-making standards.

FOREWORD

Throughout much of my career, I worked with NATO, at various times serving as commander of NATO forces in the Mediterranean, Chairman of the Joint Chiefs, and finally as U.S. ambassador in London. In all those years it was obvious to me that the greatest piece of good fortune the Alliance enjoyed was the fact that it never actually had to go to war. The strains and pressures of war would have brought all the contradictions in the Alliance to the surface and would have forced the hasty resolution of many issues that had long been kept artfully submerged. War would also have imposed on NATO the necessity to innovate—something no organization does easily. But then, eight years after the collapse of the adversary it was created to contain, NATO finally did go to war. War proved to be the difficult experience I had long foreseen, but by good luck it was a modest little war.

It was a peculiar war in the sense that it was an out-of-area operation against an opponent whose entire economy did not amount to a tenth of NATO's budget. It was, nonetheless, a protracted struggle—78 days long. Some had expected a few days of bombing to be enough to make Slobodan Milosevic relent in his campaign to drive the Moslem Kosovars out of their home villages and across the border into Albania.

It was a contentious struggle as well, with General Clark scrambling in Europe to sustain a consensus on matters as detailed as each day's target list among the nineteen governments of NATO. At the same time he was negotiating hard with his superiors in Washington for more resources and broader authority. He must have directed more attention to internal problems than he did to Milosevic.

It was also a safe war, for the NATO side particularly, as it suffered no casualties, a historic first. It was an unusually safe war for the enemy as well, since there were, by the estimates of various human rights organizations, fewer than five hundred civilian casualties. It was also a strangely civil war (in the sense of the word that suggests good manners) since it was the first war in history where, as General Clark has remarked, couples walked along the Danube and dined at sidewalk cafés while the bombardment went on around them.

It was an odd war in still other senses. It was the first war to be conducted entirely from the air, and this was odd because the stated purpose of the war—to prevent Serbian troops and police from driving innocent people from their homes—could hardly be accomplished from 10,000 feet. To stop armed men from terrorizing a village, one must put troops on the spot. This NATO was unwilling to do, so air strikes were ordered instead. These did, in time, cause Milosevic to relent (precisely why he did so remains unclear), although before he did so, he had redoubled his efforts and multiplied the number of refugees.

It will be remembered as an exceptional war in the sense of being the first ever waged and won entirely from the air. If it was, it may yet have been the threat of ground troops arriving at last that caused Milosevic to capitulate, or the restiveness of his populace, or his abandonment by his patrons, the Russians. In any case, it would be premature to draw conclusions about the new potency of air power from that odd case or to reach profound lessons of any kind from a single instance.

Already the experience of Kosovo has been contradicted, revised, and expanded by the results of the air campaigns over Afghanistan and Iraq. Operation Enduring Freedom demonstrated that air power works best when coordinated with ground forces, even if they are only a few spotters. Operation Iraqi Freedom illustrated that air power can also work well as flying artillery, and as the striking arm of small teams of Marines and special operations troops.

As this book goes to press in the fall of 2003, Syria and North Korea loom as possible targets for American air power. Recent campaigns are being carefully mined for the recommendations they may offer for future engagements. I commend these thoughtful essays to you and am particularly proud that they were produced at the U.S. Naval Academy where I graduated some years ago and where I teach today.

Admiral William J. Crowe, USN, Retired
Annapolis, July 2003

Introduction

THE PROMISE OF IMMACULATE WARFARE

Stephen D. Wrage

The author of a recent article in *Foreign Policy* got a little carried away about air power. He concluded that with precision weapons, the United States can be "the good guy" from "the old TV westerns."

> The good guy—the one in the white hat—never killed the bad guy. He shot the gun out of his hand and arrested him. Modern air power may not solve every military problem, but thanks to the innovations of the last decade, it is the weapon in the U.S. arsenal that comes closest to fulfilling that goal.[1]

His enthusiasm is understandable, but it should be countered with a little sobriety. New ways of delivering force from the air certainly offer policy makers some very good options, but they do not turn the United States into the Lone Ranger. It may turn out they were *too* good if the dramatic reports from Kosovo, Afghanistan, and Iraq make Americans believe that we can keep order and dispense justice around the world the way Matt Dillon did in Dodge—doing only what is right, saving the weak from the wicked, shooting almost magically straight, and never getting too badly hurt ourselves.

Euphoria comes easily because change has come so fast. The air campaign over Kosovo was an impressive display of precision bombing, but the campaigns over Afghanistan and Iraq went even further in the unveiling of new devices and techniques. Precision guided munitions [PGMs] drew considerable attention, but they have been in the American arsenal since the latter days of the Vietnam War. The real story in recent years has been the combination of PGMs with stealth, night, and all-weather delivery capabilities, plus satellites, drones, eavesdropping planes, and ground spotters for target acquisition, supplemented by streamlined air tasking orders and target approval procedures run with exceptional regard for avoiding unintended deaths. All these operations, like those over Kosovo, took place in the context of new ways of marshaling coalitions, managing allies, reassuring Congress, and shaping the perceptions of the American and global public so the campaign could carry forward unimpeded.

In Afghanistan, all these measures functioning together meant that loitering B-52s and AC-130 gunships could be called in to destroy a moving truck convoy twenty minutes after it was spotted. It was all so high tech that Flash Gordon, not Matt Dillon, came to mind. The public affairs officers got right to work, trying to get everyone to drop the phrase "air warfare" and start calling it "*aerospace* warfare" instead. That is to be expected. The real surprise was when they themselves, and the press and commentators, largely stopped using the phrase "collateral damage." When a classic euphemism such as that one falls out of use, it must be because it is much less needed than before. It was replaced, however, by the phrase "unintended casualty," and this became a subject of intense attention.

This is a look at air warfare taken by practitioners, by people who took part in Operations Allied Force and Enduring Freedom. All of the contributors to the volume, except for the editor and the author of the foreword, served in the campaigns over Iraq, Kosovo, and Afghanistan. As a group they are not so much skeptical as cautious and realistic. They want it recognized that the use of force is always difficult and in some respects chancy. The successes achieved in those campaigns were won only with extraordinary skill and some good fortune, and it would be hasty and overconfident to conclude, for example, that similar campaigns will be possible in the many other sites around the world as the pursuit of al Qaeda continues.

The contributors' essays make a number of points: that using force to coerce an opponent to do our will (as Slobodan Milosevic was coerced in Operation Allied Force) is always difficult, and coercion carried out from the air alone is considerably more difficult still; that warfare, even with the most skillful application of the most precise, reliable, and capable technologies, can never be perfect in its avoidance of unintended deaths and in its skirting of unintended consequences, yet the very success of recent efforts in this direction appears to be raising the expectations of many observers (including even the very experienced professional aviator quoted above) to unrealistic, perfectionist levels; that coalition warfare brings immense complications, particularly for the regional commander; and that the moral issues involved in any use of force do not vanish even if the ability to aim is vastly improved. The authors will be satisfied if people exhilarated by the apparent triumphs in Iraq, Kosovo, and Afghanistan stop and think about the extraordinary measures that were involved and the possibility that on future occasions the outcomes may be less happy.

This volume is a product of the U.S. Naval Academy in Annapolis. All of the contributors are graduates of the Academy, or instructors there, or both. All of the contributors are trained academics with advanced degrees; three of them were participants in Allied Force, two in Enduring Freedom, two in Iraqi Freedom, and one in all three campaigns, plus Northern and Southern Watch.

Major Scott A. Cooper, U.S. Marine Corps, flew EA6Bs for thirteen months over Iraq and four weeks over Kosovo. Recently he was deployed to Afghanistan, then to Iraq. He served in 2000–2001 as an International Affairs Fellow at the Council on Foreign Relations based at the Council's Washington office and studied the problems of air power and coercion. He writes from the perspective of the cockpit on the problems of evading air defenses, assuring targets are correct and legitimate, and avoiding civilian casualties.

Lieutenant Spencer Abbot flies F-18s off the Theodore Roosevelt and took part in Operation Enduring Freedom and Iraqi Freedom. He was brigade commander when he was at the Naval Academy and has been an election observer in Kosovo. He contributes the theoretical chapter on the problem of using force for coercion, for deterrence, or for compellence.

Lieutenant Derek S. Reveron, Ph.D., is an intelligence officer and a member of the Naval Academy's political science department. He served as a military and political analyst for General Clark at NATO headquarters in Belgium during Operation Allied Force. His essay highlights the extraordinary difficulties that had to be overcome to make coalition war possible, particularly the equivocal, limited, cautious war that the situation demanded and the NATO allies were willing to support.

Dr. Stephen D. Wrage is a member of the Naval Academy's political science department and the editor of the volume. He writes and teaches about the formulation of U.S. foreign policy and on issues of ethics in international affairs.

These views are their own and do not reflect the thinking of the United States military, the Navy, or the Naval Academy.

NOTE

1. Philip Meilinger, "A Matter of Precision," *Foreign Policy,* March-April 2001, 41.

Chapter 1

AIR POWER AND THE COERCIVE USE OF FORCE

Scott A. Cooper

In June 2000, Brigadier General John D. W. Corley, director of studies and analysis for the U.S. Air Force at its headquarters in Europe at Ramstein Air Base, Germany, made a bold proclamation after the publication of his 10,000-page report on Operation Allied Force. He declared, "We were able to take on Milosevic and vanquish him. We were able to meet this objective through the hard leverage of aerospace power."[1]

General Corley's remarks reflect a widely held view that air power is a cheap and easy military solution to foreign policy problems. This view takes as axioms three attractive beliefs: that air power saves the lives of U.S. soldiers on the ground, that the advanced technology of precision-guided munitions reduces collateral damage, thus making war less bloody and more morally acceptable, and that fear of this sophisticated technology coerces an enemy to do our will. In sum, air power has come to seem like a silver bullet—a surefire, low-risk, high-performance instrument to be used with little cost yet to great effect.

Although Corley is right to praise the consistent and dependable performance of the U.S. military, his unreserved endorsement of air power should not be taken at face value. It is dangerous for any strategic thinker or policy maker to believe that air power used alone can vanquish an enemy. Operation Allied Force was a successful test of NATO's will and cohesion. Its lessons, however, are not found in its victory, but in its problems and paradoxes.

COERCION AND CAPITULATION

At the outset of Allied Force, policy makers expected they could use air power to stem the flow of refugees out of Kosovo. They ought to have known and in fact soon learned that air power had little or no impact on the Serbs' efforts to expel Kosovars from their homeland, except perhaps to hasten them. Air power was chosen because it was something that could be done, and done quickly, even though it hardly related to the situation on the ground.

Many in the defense community, including those who flew in Operation Allied Force, were acutely aware of the limits of air power. After the week or two of bombing, the pilots and air crews were asking the same question as many editorial writers: "What is air power doing to stem the flow of refugees?" Aviators quickly came to understand that their mission was not to curb the refugee tide but to destabilize Milosevic's regime. They realized they were being employed in preference to ground troops simply because the coalition arrayed against Milosevic could not agree on and commit to costlier, riskier measures. Milosevic's eventual capitulation surprised those who flew in the war as much as it did the general public, and many fliers did not believe that the 11 weeks of bombing had directly caused the retreat of Serb forces from Kosovo. No one has yet explained why Milosevic relented and withdrew, and until we know why he did so, neither General Corley nor anyone else can convincingly declare it was "aerospace power" alone that "vanquished" him. Operation Allied Force makes it tempting to reach triumphant conclusions about the irresistibility of American air power, but until we can confidently say why air power unaccompanied by ground forces worked to coerce an enemy regime, we cannot be confident that it will work again.

Operation Desert Storm, the air campaign against Iraq, makes such overconfidence equally tempting, but like Allied Force, Desert Storm had its odd, unlikely to be repeated characteristics. Why, for example, did Saddam choose to fight the coalition symmetrically, and would another dictator, seeing the dismal way Saddam's forces fared, make the same mistake? Why did Saddam not use chemical weapons against Israeli cities, and who could be confident that he would not act differently in another war? Moreover, when will American forces be so lucky again as to have a battlefield so clear, so visible, so open, yet so well defined and contained? It was largely a desert war, and war in the desert, like war at sea, makes it easy to aim exclusively at combatants and to use force freely but without unintended effects. Air power will seldom be so easy to use.

AIR POWER AND THE COERCIVE USE OF FORCE

In Kosovo as in Iraq, luck favored the alliance. NATO was lucky first in that it was never actively opposed by the Serbian Integrated Air Defenses (IADS). The Serbian military conserved its resources (surface-to-air missiles and anti-aircraft artillery, but also tanks, armored personnel carriers, and other heavy equipment) and waited, hoping the alliance would fold. This strategy meant that NATO's opposition was never a true competitor, much less a peer competitor. More aircraft would have been lost had the Serbian IADS more aggressively confronted NATO.

Second, dramatic scenes of 800,000 Albanian refugees flooding into neighboring countries strengthened NATO's resolve. Had it not been for the humanitarian crisis in Kosovo, in all likelihood the length and scope of the campaign would have given rise to more significant opposition from U.S. and allied policy makers than it did. In April 1999, during the height of the conflict, at the fiftieth anniversary celebration of NATO, allied leaders codified their resolve and pledged to bring the mission to its successful conclusion. It was a nice piece of luck that such a highly symbolic moment for reaffirmation of purpose should have fallen in the midst of the campaign, yet so great were the misgivings in Washington about Allied Force that the NATO commander, General Wesley Clark, USA, was initially excluded from the affair.[2]

Lastly, in Operation Allied Force NATO was fortunate because Operation Horseshoe did not call for genocide. Had Milosevic ordered that the Kosovar Albanians be systematically killed rather than forced out of Kosovo, the severe limits of air power and of U.S. military power in general would quickly have become obvious, for it would have taken a significant ground force to expel Serbian forces from Kosovo and stop a genocide. It is unlikely that the United States would have stepped away from a massacre of thousands upon thousands of Kosovar Albanians, but the mission would have been challenging and costly.

When Operation Allied Force began, most in the military heard or read transcripts of President Clinton's speeches to the U.S. people, in which he proclaimed:

> Our strikes have three objectives: First, to demonstrate the seriousness of NATO's opposition to aggression and its support for peace. Second, to deter President Milosevic from continuing and escalating his attacks on helpless civilians by imposing a price for those attacks. And, third, if necessary, to damage Serbia's capacity to wage war against Kosovo in the future by seriously diminishing its military capabilities.[3]

Charged with this mission, aviators, typically sarcastic, joked among themselves in the ready room, "Now, how are we going to stop this ethnic cleansing from a jet? If Serbs line up Albanians and start shooting them, what are we going to do? Maybe we could fly over, drop a flare, and yell at them to stop shooting!" Such flippant and cynical comments were born from the realities of the battlefield the aviators faced. A pilot in an airplane does not easily grasp the logic of a landscape beneath him, no matter how accurate his targeting pod or his radar.

The crew of an aircraft can be expected to bomb a fixed target for which they have a satellite image, even with significant cloud cover. Technology allows aviators to determine which part of a building to hit for maximum impact and even what damage can be expected on the surrounding structures, all while avoiding air defenses as simple as a man with a shoulder-fired missile or as complex as a fiber-optically linked, multitiered radar operations sector. To be expected to stop lightly armed military police from killing unarmed civilians, however, mismatches the mission and the means.

In the celebratory aftermath of achieving these objectives, it is important for the United States to remember to match a particular use of military force to its foreign policy objectives, and not depend solely on victory through airpower.

"The distinction between the power to hurt and the power to seize or hold forcibly is important in modern war."[4] Air power offers only the ability to hurt, not the ability to seize or to hold. If air power is the only tool, the critical question becomes how much damage the opposing side is willing to incur, and the hope to "vanquish" the enemy may have to be deferred. Initiative is ceded to the targeted side, for it is the opposition's reaction to the pain inflicted that determines the length and effectiveness of the bombing.

In cases where the enemy has a much greater stake in the outcome, coercion is a risky and uncertain course of action, as was evident in Operation Allied Force. NATO's overconfidence in the utility of air power was deflated when the flow of Kosovar Albanian refugees did not slow. Only after eleven weeks of bombing did NATO induce Serbia's capitulation and the entry of a NATO-led occupation force into Kosovo. The splendid success of Operation Allied Force was its facilitation of the arrival of ground forces without the struggle of fighting for the territory. Credit for the operation's success does not lie exclusively with the coercive use of air power but rather with the confluence of several factors: the threat of a ground invasion, the withdrawal of Russian diplomatic support, and the increasing damage wrought by the air campaign.

As Thomas Schelling wrote in *Arms and Influence,* "Compellence has to be definite: we move, and you must get out of the way."[5] We choose our target to make the enemy move, and we set our level of force to extort a compliant response, but once we make our move, we must wait for the enemy's response, and it is never possible to say with confidence what that response will be. Once force is committed, if the adversary's behavior does not change, our only options are escalation or admission of failure.

BETTER MODELS

A superficial consideration of Desert Storm and Allied Force will tempt policy makers to put too much confidence in the use of air power alone to coerce an adversary. Operation Deliberate Force, the series of air strikes against Bosnian Serbs in 1995, offers clearer lessons, and these lessons are reinforced by the unfolding experience of Operation Enduring Freedom in Afghanistan.

By the summer of 1995, Bosnian Serbs and their allies from Serbia proper had been "cleansing" Bosnia of its Muslim population for over three years. They had seized almost three-quarters of the country, and the city of Srebrenica had been under siege for most of that time. The Serbs were shelling Bihac, an enclave four miles by two containing up to 200,000 Muslim refugees. NATO planes with American pilots struck at airfields used by the Serbs, and the Bosnian Serb leader, Radovan Karadzic, responded by taking United Nations Protection Force members hostage and threatening to treat them "not as peacekeepers but as enemies."[6] To that point coercive bombing clearly had backfired.

Violence escalated with attacks by Bosnian Serbs on two of the safe areas: Gorazde and Srebrenica. Srebrenica at last fell on July 11, 1995, and mass deportations and executions followed. Finally, on August 28 the Serbs shelled a Sarajevo marketplace, killing nearly forty people and emboldening—indeed practically requiring—NATO to launch Operation Deliberate Force.

The operation ultimately consisted of 293 aircraft from eight NATO countries flying 3,515 sorties over the course of two weeks. The stated aim of the campaign was to coerce the Bosnian Serbs to cease their military offensive against the Bosnian Croats and to accept a peace deal, but also to shift the military balance in favor of the Bosnian Croats. Initially, attacks focused exclusively on Serbian command and control sites, surface-to-air missile and anti-aircraft artillery sites, and supporting radar and communications facilities. In other words, they concentrated on the integrated air defenses in Bosnia. Later attacks targeted ammunition dumps, artillery positions, communications links, supply storage areas, and finally some key bridges used to support Bosnian Serb military operations. The targets were carefully selected to avoid destroying the entire Serbian infrastructure. The air strikes, however, never did alter the military balance. The Croatian ground offensive accomplished that.

The strikes were even suspended for two days on September 2, to test Serbian compliance to NATO's demands, reminiscent of Operation Linebacker II and the Paris peace negotiations with North Vietnam in 1972. NATO involved itself in what Schelling would call an exercise in coercion by manipulation of the enemy's risk. Air Force chief of staff General Ronald Fogleman emphasized this in a letter to the editor in the *Wall Street Journal.* He stated that the aim of the bombing was "not to defeat the Serbs, but simply to relieve the siege of UN safe areas and gain compliance with UN mandates and thus facilitate ongoing negotiations to end the fighting."[7]

Ultimately, NATO secured safe areas, removed offensive heavy weapons, and reopened airport and road access to Sarajevo. Perhaps most importantly, the campaign's success laid the foundation for the Dayton peace accords, signed in Paris on December 14.

The lesson drawn from Deliberate Force seems to have been the vindication of air power's effectiveness as a coercive instrument, while forgetting the enormous risks involved in such a strategy. Ambassador Richard Holbrooke, the primary negotiator of the Dayton agreement, observed that the air campaign had made a "huge difference" in helping to bring about an acceptable outcome.[8] Yet it was truly a fortuitous confluence of events that resulted in the Dayton accords. Driving the Bosnian Serbs from the Krajina region were two military operations: the NATO air campaign and the Bosnian Croatian ground offensive. The Croatian offensive resulted in a close approximation of the nearly equal division of Bosnia in 1994, thus negating many of the Bosnian Serb gains of the previous year. Moreover, Milosevic abandoned the Bosnian Serbs at Dayton to secure Serbia proper and Yugoslavia. Thus, air power was only part of a series of circumstances that drove the Bosnian Serbs to negotiate a settlement.

Lost in the celebration over Dayton was a serious analysis of the effects of coercion. Throughout Operation Deliberate Force, the power to deescalate the conflict was in the hands of the Bosnian Serbs, whose pain threshold was the determining factor in negotiating a settlement. What if the Bosnian Serbs had accelerated the violence and continued the siege on Sarajevo and Srebrenica? What if Slobodan Milosevic had supported the Bosnian Serb offensive with equipment or manpower? The NATO allies may have been in the same position as during Operation Allied Force: executing what many characterized as a "bomb and pray" strategy.

LIMITS OF COERCION

Operation Allied Force was an extremely destructive use of force, but an extraordinarily well aimed one. The ability to aim so precisely raises anew ancient issues from *jus in bello,* the judgments regarding how force may be used in the conduct of war. Kosovo may represent an important turning point in the ability to direct force solely at legitimate targets and so may have changed what Michael Walzer calls "the moral reality of war."[9]

In World War II, the United States and its allies inflicted as much damage on Japan and Germany as they were capable of inflicting. In his book, *Danger and Survival,* McGeorge Bundy portrays the decision to bomb Hiroshima as one in which no one in the upper reaches of the U.S. government raised the question of civilian casualties. By contrast, during the Gulf War, enormous effort was made to reduce circumstances in which civilians would be killed. No doubt there has been a learning curve from World War II to Kosovo. Today it is unacceptable, both to the U.S. public and among the world's decision makers, to pursue a policy of purposeful and intentional killing of civilians.

Many reasons exist for the strong consensus on limiting the destructiveness of force, but two stand out when considering the use of air power. First, most U.S. military commitments in the last fifty years did not constitute a direct threat to the United States, and therefore did not merit extraordinary use of force. Compare this to the threats faced during World War II, which for our British allies constituted what Walzer and others have called "a supreme emergency" and for the United States required a total mobilization of five years' duration. Thus, the greater the direct threat, the easier it is to do things that would otherwise be considered reprehensible. In altruistic interventions such as that in Kosovo, U.S. tolerance for civilian deaths diminishes.

The second reason we have come to believe in limiting the destructiveness of force is the myth of perfect precision. Broadcast images of laser-guided bombs hitting the correct window of a building taught the public to expect flawless warfare free of unintended deaths, but accidents will still occur and civilian loss can be catastrophic.

The promise of a transformed moral reality of war, coupled with U.S. sensitivity to the loss of life, directly affected how Operation Allied Force was conducted. Minimal collateral damage made the war a success in the eyes of the general public, but these factors limited the alliance's ability to conduct a rapid and effective coercive air campaign against Serb forces. Although moral considerations constitute a valid limit on coercive strategy, these mean practical limitations that can get in the way of a successful campaign.

As a result, the central story of Operation Allied Force from the point of view of headquarters was the struggle to determine the enemy's center of gravity and choose which targets to hit. Those aviators flying the operation's missions were acutely aware of the disagreement between Supreme Allied Commander General Wesley Clark and Joint Forces Air Component Commander Lieutenant General Martin Short about Yugoslavia's strategic center of gravity.

General Short vocally disapproved of the restraints placed on targets during the campaign. After the war, in a statement before the Senate Armed Services Committee, he reiterated his point: "I believe the way to stop ethnic cleansing was to go at the heart of the leadership, and put a dagger in that heart as rapidly and as decisively as possible."[10] General Clark, on the other hand, believed that the air campaign should focus on the fielded forces in Kosovo who were conducting the ethnic cleansing campaign, and thus contended that the B-52s bombing the Serb forces in Kosovo would have better effect than the strikes on targets in Belgrade.

In a now well known exchange during one of the daily video teleconferences, General Short expressed satisfaction that, at last, NATO warplanes were about to strike the Serbian special police headquarters in downtown Belgrade. "This is the jewel in the crown," Short said.

"To me, the jewel in the crown is when those B-52s rumble across Kosovo," replied Clark.

"You and I have known for weeks that we have different jewelers," said Short.

"My jeweler outranks yours," said Clark.[11]

The disagreement between Clark and Short about the Yugoslav center of gravity is instructive about the use of air power for the future, especially if one advocates the sort of bombing campaign Short does. The theory goes that these targets are the Achilles' heel of the enemy; that, if destroyed, the central leadership will be isolated and the enemy's military will collapse under light military pressure due to a lack of guidance. In Operation Allied Force, these targets were Milosevic, his cronies, and the industries and buildings they personally valued, such as counterintelligence, security forces, loyal military units, and the related communications facilities.

Such a strategy has historically been ineffective.[12] The only successful case of wartime assassination by military forces was that of Japanese admiral Isoroku Yamamoto in World War II, and it had no effect on the outcome of the war. The operation in Panama in 1989 showed how difficult it is to find an enemy leader, when it took U.S. troops days to apprehend General Manuel Noriega. Also, efforts to target Saddam Hussein in both Gulf Wars proved unsuccessful. Short's type of strategy also sacrifices much of the moral authority we bring to a conflict, not to mention the presidential directive prohibiting the assassination of foreign political leaders. Furthermore, accomplishing the mission does not guarantee the end of conflict or a predictable succession. During Operation Allied Force, Serbs were hostile to the bombing, no matter how much they despised Milosevic. Images of Serb civilians wearing T-shirts with targets on them should have warned us that the strikes were not likely to end Milosevic's political control of Yugoslavia, or break the will of the people. Indeed, it was an internal democratic movement that led to the downfall of Milosevic, not an air campaign.

Short's call to target Yugoslavia's central leadership and communications facilities focused on one possible good—a quick capitulation—and ignored many certain evils. Short's plan to "go downtown" carried with it a huge moral cost. Had we turned out the lights in Belgrade and eliminated the leadership, we would have killed many civilians and increased the likelihood of more incidents akin to the bombing of the Chinese embassy. This is not to say that such a campaign could not have been waged successfully, but to do so would have wreaked more havoc than allied leaders were willing to accept and would have fractured the alliance. This type of campaign also risks a disproportionate use of force, with harm done outweighing the military utility achieved. Indeed, coercion carries with it the possibility of undermining the larger good we are pursuing in the campaign. As Elliot Abrams cautioned, the picture "of a superpower willing to bomb but not to fight, willing to inflict a tremendous amount of pain on others to avoid the slightest risk to itself...that is a picture that should repel us."[13]

If we assume, however, that General Clark was correct in his view of the Yugoslav center of gravity being the fielded Serb forces in Kosovo, more effort should have been placed on these units. Such a focus, however, again demonstrates the limits of air power. Fielded forces are hard to hit. Air power relies exclusively on accurate and timely intelligence for effectiveness. Such intelligence is often hindered by weather, terrain, camouflage, and concealment, among other factors that have always influenced military operations.

A costly demonstration of these challenges was the accidental bombing of the Chinese Embassy. The Gulf War spoiled us, because the desert provides a much easier intelligence and targeting challenge than the mountains and forests of Yugoslavia. Serb forces went to great lengths to hide their equipment, avoid overhead detection, and disperse their troops.

Clark and Short's disagreement, coupled with their joint commitment to using airpower, evolved into a dual-track strategy in the first week of the air campaign, with nighttime attacks on preplanned targets in Kosovo and Serbia proper, and daytime "kill boxes" in which pilots would loiter above Kosovo looking for targets of opportunity, that is, the Serbian Third Army. The daytime sorties proved to be a Herculean effort with few measurable successes. The Serbian fielded forces operated primarily out of Fiats, Volkswagens, and other civilian vehicles, not tanks and armored personnel carriers. The difficulty of targeting became apparent early in the conflict in the unfortunate attack near Dakovica on a refugee convoy presumed to be a troop movement of Serb forces.

Pilots faced severe targeting problems from the first day they flew the daylight kill boxes. Dozens of aircraft would fly into Kosovo looking for "targets of opportunity," or Serb military forces in Kosovo, only to find the vast majority of the time that no targets could be found. Serbs were not in the open where the aviators could detect them. Moreover, without troops on the ground to act as forward air controllers to direct pilots to the location of enemy forces, the likelihood of finding a target was very low. Instead, bombs would be dropped on relatively insignificant "dump" targets, such as roads, or brought back to base to be used another day.

Aviators were acutely aware of the impact of even one misguided bomb or missile. The United States possesses powerful technology, but that is no guarantee against grievous mistakes, as was the case when a bridge over the Danube was accidentally bombed just as a passenger train was crossing it. Certainly, precision weapons allow fewer aircraft to destroy more targets, but war is always destructive, no matter how precise the weapon.

A focus on the fielded forces in Kosovo also lent itself to an ill-advised dependence on quantifying the campaign. The raucous debate over the number of Serb military targets destroyed by allied aircraft detracted from the overall aims of the mission and undermined the concept of coercive strategy. In the end, the dueling opinions of Clark and Short were resolved into a single targeting policy that contributed to the end of making Milosevic back down.

LESSONS FOR THE FUTURE OF AIR POWER

Operation Allied Force offers many important lessons when considering the use of force for the future. First, although Allied Force was not a punitive strike, punitive strikes are political, and air power is a relevant tool to make a political statement. Such actions carry much less risk than using air power as a coercive tool. Generally, punitive strikes serve domestic interests and rarely have strategic effect. Often the determination of "how much is enough" is driven by the interpretation of public opinion. From the raid on Libya in 1986 to the cruise missile attacks on Iraq in 1993 to the 72 hours of air strikes against Iraq in December 1998, such operations have satisfied political concerns, answering the call, "Do something!" They do, however, result in little change to the strategic realities on the ground.

Second, Operation Allied Force taught the military what potential enemies have learned in the decade since Operation Desert Storm. Aviators who flew in both campaigns faced the pressing challenge of neutralizing the enemy's extensive network of lethal surface-to-air (SAM) missiles. The Iraqi air defense forces paid a heavy price in the beginning hours of Operation Desert Storm, with hundreds of high-speed antiradiation missiles (HARMs) raining down on detected radars. Iraqi radar operators were intimidated into operating in less-than-optimal modes for the rest of the war, out of fear of antiradiation missiles. This lesson was not lost on the Serbs, who kept most of their SAMs hidden and the SAM radars not emitting, deliberately husbanding their assets out of respect for, or fear of, the allied antiradiation missiles. Throughout Operation Allied Force, a credible SAM threat persisted, but the use of HARMs during Allied Force proved to be much less effective than in Operation Desert Storm. In the last ten years the Serb military and other potential enemies learned a great deal, potentially limiting future dominance of airspace, especially given this surface-to-air threat.

Third, Operation Allied Force should serve as a warning. After Operation Desert Storm, Eliot Cohen warned of the seductiveness of air power, comparing it to modern teenage romance. It offers political leaders a chance for "gratification without commitment."[14] Air power is often viewed as the universal remedy when diplomatic means are exhausted. Bombing is not a strategy, however, and coercive bombing campaigns have a spotty record. Policy makers should be mindful that air power carries enormous risks and costs.

AMERICAN LEADERSHIP AND THE USE OF FORCE

Aviators who flew in Operation Allied Force were probably still safer flying over Kosovo getting shot at by surface-to-air missiles than the ground troops carrying M-16s in Pristina as part of the NATO-led occupation force. The heavy lifting and most dangerous part of the operation in Kosovo was and is being carried out by the troops on the ground. The Kosovo peacekeeping mission is an extension of Operation Allied Force, and the United States and its allies should not believe that they have executed an operation without costs.

Humanitarian uses of the military, such as in Kosovo, are not likely to abate in the near future. As retired Marine General Anthony Zinni, the former commander in chief of Central Command, stated in a speech to the U.S. Naval Institute in March 2000:

> [W]e're going to be doing things like humanitarian operations, consequence management, peacekeeping, and peace enforcement... And somewhere else along the line we may get stuck with putting a U.S. battalion in place on the Golan Heights, embedded in a weird, screwed-up chain of command. And do you know what? We're going to bitch and moan about it. We're going to dust off the Weinberger Doctrine and the Powell Doctrine and throw them in the face of our civilian leadership.... [M]ore and more U.S. military men and women are going to be involved in vague, confusing military actions—heavily overlaid with political, humanitarian, and economic

considerations. And representing the United States—the Big Guy with the most formidable presence in the area—they will have to deal with each messy situation and pull everything together.[15]

General Zinni touched on an important part of the military's role as it has evolved over the past decade. The military is not well designed for nation-building and other missions with which it has been tasked. Nonetheless, the military will continue to face these challenges because it is the only entity of the U.S. government capable of carrying out such nation-building missions. General Clark, reflecting on the post-conflict mission in Kosovo, recently lamented that he needed Robert Komer and CORDS (CORDS was the Civil Operations and Revolutionary Development Support—a civilian pacification organization in Vietnam headed by Komer that worked side-by-side with the U.S. military) in Kosovo to handle the peacekeeping and nation-building tasks of the mission. Until we rebuild our nonmilitary diplomatic tools, however, the military may be the only U.S. organization that can facilitate change in postconflict situations.

To avoid shirking its responsibilities of leadership, the United States must recognize its extraordinary role in the world and understand and accept the ambiguous and complex political circumstances in which our military will be called on to operate. When deciding to use force, the United States must be prepared for the possible consequences and enter into such commitments without any illusions regarding what force can do, what harbingers of success or failure may accompany such commitments, and what the costs may be. We cannot assume that the situation on the ground will lend itself to symmetrical battles or situations where coercive strategies will work.

Since World War II, U.S. policy makers have sought military solutions to political problems. The president has been tempted to seek help from the military because it has been so readily available and possesses an impressive track record. No adversary in recent years, it seemed, could stand against the American military. This is the exact opposite of the attitudes in the wake of the failure in Vietnam, when it was believed that the United States could accomplish relatively little by force of arms. Today the pendulum has swung again to an overzealous confidence in the United States' ability to accomplish its goals by military means.

When it comes to the use of force, often the question is not whether to use force, but how. Military force is going to remain relevant to a variety of tasks, and the U.S. military will be ordered to carry the lion's share of any demanding military operation. Nevertheless, the extraordinary success of Operations Desert Storm and Allied Force, as well as the remarkable achievement in Afghanistan, carry with them a dangerous confidence. Our success should be celebrated, but we should be wary not to anticipate the kind of optimal war-fighting circumstances that we found in the Gulf or the still unexplained willingness to relent that we found in Milosevic. Asymmetric competitors will stun us with new methods and measures, just as they did on September 11th. To remain a great power we will have to overcome the fog and friction of war and adapt to the inventive efforts of adversaries bent on feeling out our points of vulnerability. It is not enough simply to possess overwhelming firepower. Successful execution of a coercive strategy demands that we possess the airplanes and weapons to do the mission, but also maintain a realistic understanding of what we can hope to accomplish.

NOTES

1. Technical Sergeant Joe Bela, U.S. Air Forces in Europe Public Affairs, Ramstein Air Base, Germany (Air Force Public Affairs, 2 August 2000, http://www.af.mil/news/Aug2000/n20000802_001154.html). Accessed 13 February 2001.

2. David Halberstam, *War in a Time of Peace* (New York: Scribner, 2001), 469.

3. Statement by the President on Kosovo (White House, Office of the Press Secretary, March 24, 1999).

4. Thomas Schelling, *Arms and Influence* (New Haven, Conn.: Yale University Press, 1966), 6.

5. Schelling, *Arms and Influence,* 72.

6. Halberstam, *War in a Time of Peace,* 285.

7. General Ronald R. Fogleman, USAF, "What Air Power Can Do in Bosnia," letter to the editor, *Wall Street Journal,* 11 October 1995.

8. Richard Holbrooke, *To End a War* (New York: Random House, 1998), 104.

9. Michael Walzer, *Just and Unjust Wars* (New York: Basic Books, 1977).

10. Lieutenant General Michael C. Short, testimony, Senate Armed Services Committee, *To Receive Testimony on the Lessons Learned from the Military Operations Conducted As Part of Operation Allied Force and Associated Relief Operations, with Respect to Kosovo.* 106th Cong., 1st sess., 21 October 1999, 399.

11. Dana Priest, "The Battle inside Headquarters," *The Washington Post,* 21 September 1999, A1.

12. The classic case of its ineffectiveness is Robert Pape, *Bombing to Win* (Ithaca, N.Y.: Cornell University Press, 1996).

13. Elliott Abrams, "Just War. Just Means?" *National Review,* 28 June 1999, 18.

14. Eliot A. Cohen, "The Mystique of U.S. Air Power," *Foreign Affairs,* January/February 1994, 109.

15. General Anthony Zinni, USMC. "A Commander's Reflections," *Proceedings* 126/7/1169 (September 2000): 61.

Chapter 2

AIR POWER STRATEGY AND THE PROBLEM OF COERCION

Spencer Abbot

During the 1990s, U.S. policy makers responded to numerous crises by either threatening the use of air power, or by integrating air power strategies within broader diplomatic and economic efforts to modify the decision-making processes of targeted states and organizations. The seventy-eight-day air campaign that the U.S. waged against the Serbian government in 1999, Operation Allied Force, is a product of evolving strategic thought within the U.S. military, and the U.S. government more broadly, regarding the relative merits of various strategies for the employment of the United States' formidable air power capabilities. Over the past decade policy makers and academics alike have devoted increased attention to the study and implementation of air power strategies in the broader context of coercion and coercive diplomacy, and those efforts have directly impacted the planning and conduct of air power employment in recent years, from air campaigns in Bosnia and Serbia/Kosovo, and more recently in the use of military force in Afghanistan and Iraq.

CONTROL FROM ABOVE?

The primary dilemma facing policy makers at the outset of the NATO air operations against Serbia in March 1999 was whether an air campaign could successfully coerce Milosevic without the threat or employment of a ground campaign, and if not, what then? What tools did those policy makers and senior military leaders have available at the time to assess air power's potential to successfully force Milosevic's hand, and thereby achieve the United States' stated policy objectives? Air power, even in its most precise manifestations, is an enormously powerful but fundamentally blunt force. Over the past decade, policy makers and military strategists have found new ways to couple air power with strategies of deterrence, persuasion, and coercion, which describe the manner through which force, or threats of force, can be used to shape the decision-making processes of targeted states or organizations. Ultimately, any coercive strategy seeks to shape events and decisions occurring on the ground, and the most ambitious strategies seek not just to impact but to exert a degree of control over those events. Clearly, air power is one of the United States' comparative military strengths and has thus been frequently employed over the past decade in pursuit of U.S. and NATO strategic objectives. However, the potential for air power to impact targeted decision-making processes precisely and predictably enough to move beyond coercion to actual control has not yet been clearly proven or established, and policy makers must guard against overconfidence in such expectations when embarking on air campaigns or employment of airborne strike assets.

Predicting the success of air power as a coercive instrument has proved to be particularly difficult. Unplanned secondary and ancillary effects abound, and the law of unintended consequences seems ever present when air power strategies are employed. The inherent nature and characteristics of air power as a strategic instrument—its relative bluntness, indiscriminateness, and distance as compared to direct action by ground forces—render elusive any exact prediction of its effects. Moreover, discerning the exact impact of an air campaign on the decision-making process of adversaries seems always a source of great controversy. Yet air power has frequently proved the policy tool of choice in recent years in attempts to coerce nonpeer adversaries. The technological superiority of the United States in the aerospace field, along with the relative invulnerability to casualties offered by the use of air and space rather than ground-based assets, creates a tempting option for decision makers seeking coercive leverage that will at the same time be domestically palatable. Air power does possess some unique attributes that provide the opportunity for flexible utilization of force in support of particular political ends while maintaining a relatively low vulnerability to casualties. Furthermore, recent innovations in aerospace technology have enabled the development of new strategies governing its use as a coercive instrument, a

AIR POWER STRATEGY AND THE PROBLEM OF COERCION 23

development that some analysts have placed under the rubric "revolution in military affairs."

During the 1990s, as air power emerged as the coercive tool of choice for policy makers in numerous crises, much effort was expended, quite fruitfully, by air power strategists seeking to adapt new technological capabilities to an altered geostrategic arena. Modern technologies, both within the aerospace field and in the areas of communication and information processing, have altered the modalities through which air power is employed, as well as the efficacy of various strategies for its use. Strategies for the employment of air power must ultimately be viewed through the lens of the shaping events and decisions on the ground. Yet, the use of air power may or may not be coupled with the use of ground power, or the threat of future application of ground power. The strategies and methods through which air power has been used in attempts to influence the decision-making processes of other states and organizations during the past decade are reflective of theories regarding the processes and mechanics of coercion, which for many years were overshadowed by deterrence theory focusing on nuclear confrontation between the United States and the Soviet Union. During the past decade, the concepts and dynamics of coercion and coercive diplomacy have come to the forefront of both international relations theory and practice. A primary means for the operationalization of these strategies has been the threat and use of air power.

THE LATE-BLOOMING DEVELOPMENT OF COERCIVE THEORY

During the Cold War, the strategy of deterrence was intellectually predominant because of the political and strategic environment prevailing at the time. The looming threat of nuclear confrontation between the superpowers led deterrence theorists to presume that any direct use of military force by one superpower against the other would result in an unacceptable risk of wide-scale conflict and nuclear apocalypse. Deterrence theory demanded an indirect, defensive strategic posture that would minimize the potential for direct confrontation and provide for system stability. Deterrence theory thus favored the strategy of containment, which provided the primary intellectual basis for U.S. Cold War foreign policy.

Less attention, understandably, was devoted to the study of military coercion and the coupling of threats and diplomacy in pursuit of policy objectives.[1] During the Cold War, theoretical development of the concepts of coercion and coercive diplomacy was largely confined to the work of Thomas Schelling and Alexander George.[2] Although most efforts to manage contemporary crises generally incorporate both coercive and deterrent strategies, the interrelationship between strategies of deterrence and coercion when applied concurrently has been the subject of relatively little study.[3] A recent attempt to address this shortcoming is the suggestion of "strategic coercion" as an umbrella concept comprising strategies of both deterrence and coercion. Lawrence Freedman defines strategic coercion as "the deliberate and purposive use of overt threats to influence another's strategic choices."[4]

Freedman's work has begun to break down the intellectual firewall between theories of deterrence and coercion that was a Cold War byproduct, an artificial divide in the logic of coercion and persuasion that limited our understanding of how to best tailor strategy to achieve policy objectives through the threat and use of nonnuclear military force. In 1946, as the atomic age dawned, deterrence theorist Bernard Brodie observed, "Thus far the chief purpose of our military establishment has been to win wars. From now on its chief purpose must be to avert them. It can have almost no other useful purpose."[5] Brodie's statement highlights the extent to which deterrence was separated from coercion as the primary basis for Cold War military strategy. Yet in the post–Cold War, coercive strategies have been used frequently by U.S. and NATO policy makers in efforts to stabilize political or humanitarian crises, as well as in attempts to control the decisions and actions of other actors. The study of the interaction between deterrent and coercive strategies when used in consonance has been a neglected yet important vein of research, as any threat-based effort to control the actions of an adversary must necessarily involve a deterrent component—to inhibit undesired actions—and a coercive component—to compel desired actions. Strategies designed to control the decisions and actions of other actors may seek a very high level of control, requiring substantial resources and extensive feedback mechanisms for evaluating and reacting to movements by or within the targeted actor. Control-seeking efforts may as well have much more limited aims, calling for modified strategies and a lesser commitment of resources. Any prediction of the ability of air power to persuade, coerce, and control must reflect the logic of both deterrence and coercion, two historically dissonant yet related veins of study.

COUPLING DETERRENCE AND COERCION

Deterrence can be defined as the utilization of implicit or explicit contingent threats to persuade an actor to refrain from a specific course of action based on the actor's calculation of its rational self-interest. The concept of deterrence is an outgrowth of the realist notion of a "balance of power." Deterrence theory emphasizes the role of rationality in the conduct of foreign policy and presumes the calculus of an actor's self-interest to be discernable and, to some degree, predictable.[6] To be successful, a deterrent threat must be credible and sufficient. The credibility of a deterrent threat is itself a factor of two components, the capability of an actor to carry out a given threat, and its will to do so if required.[7]

Coercion is an effort to cause an actor to undertake a given course of action by manipulating the actor's incentive structure and decision-making processes through either the threat or the use of force.[8] This definition of coercion would comprise Thomas Schelling's concept of compellence, as well as Alexander George's idea of coercive diplomacy.[9] George's study of coercive diplomacy, published during the later stages of the Vietnam War and then revised in 1994 to include analysis of several new case studies including the Gulf War, exists with Schelling's *Arms and Influence* as the most influential study of the process of military coercion. Yet, there are shortcomings in George's classification of the concept of coercive diplomacy.

George posits that his formulation of the concept of coercive diplomacy is restricted to "defensive" uses, which he defines as "efforts to persuade an opponent to stop or reverse an action." The defensive nature of the strategy of coercive diplomacy would thus rule out "offensive" uses of coercive strategy, which he defines as the employment of threats "to persuade a victim to give up something of value without putting up resistance."[10] Yet George maintains that, in addition to persuading an opponent to stop short of a goal, or to undo an action already carried out, the use of threats to force "a change in the composition of the adversary's government or in the nature of the regime" could be considered a defensive use of coercive diplomacy.[11]

Few strategic goals have a greater effect on the opposing regime's will to resist than efforts to displace the leadership of opponent states. Undermining Saddam Hussein's regime in Iraq was a long-standing strategic goal of U.S. policy, and deliberate U.S. efforts materially contributed to Slobodan Milosevic's unseating through Serbian democratic process and his ultimate handover to the International Criminal Tribunal at The Hague. Peter Viggo Jakobsen has noted that such a goal "can hardly be called a defensive objective as it goes way beyond the restoration of the status quo ante."[12] Lawrence Freedman goes further and suggests that demanding a change in the composition of a regime is a fundamental challenge to sovereignty and therefore an offensive demand. Additionally, he holds that since George himself establishes a spectrum of strategies of coercive diplomacy rather than a clear dichotomy, George's "insistence that coercive diplomacy is defensive is neither tenable in practice nor useful analytically."[13]

In his 1998 book *Western Use of Coercive Diplomacy after the Cold War,* Jakobsen presents a new and slightly varied taxonomy linking coercive diplomacy, coercion, and deterrence. He incorporates Freedman's notion of "strategic coercion" as an umbrella concept comprising both deterrence and compellence, and considers coercive diplomacy and "blackmail" to be further subsets under the subheading of compellence.[14] Secondly, Jakobsen presents a useful summary of the five conditions identified by Thomas Schelling for success of a compellent strategy:

- The threat conveyed must be sufficiently potent to convince the adversary that noncompliance is too costly.
- The threat must be perceived as credible by the adversary, that is, he must be convinced that the coercer has the will and the capability to execute it in case of noncompliance.
- The adversary must be given time to comply with the demand.
- The coercer must assure the adversary that compliance will not lead to more demands in the future.
- The conflict must not be perceived as zero-sum. A degree of common interest in avoiding full-scale war must exist. Each side must be persuaded that it can gain more by bargaining than by trying unilaterally to take what it wants by force.[15]

CREDIBILITY: CAPABILITY VERSUS WILL

A compellent threat must be perceived as credible, which again is a function of the coercer's will and capability, according to similar logic as that describing credibility in a deterrence-based strategy. Capability is a measure of the resources that could potentially be devoted to implementing a given strategy or threat. The distribution of relative capabilities in the post–Cold War international system endows the United States with a profound advantage in the military arena, especially when coupled with the capabilities of its NATO allies.

All of the conflicts entered into by the United States over the past ten years have involved the use of military force or threats of potential military action by the United States against adversaries that have possessed significantly lesser military capabilities. Hence, in the implementation of coercive or deterrent strategies by the United States, capability as a relative variable has been clearly established in the favor of the United States, though the question of the capability of air power assets to unilaterally affect decision-making processes has been a substantial source of debate. Therefore will, or perceived will, exists as a primary variable affecting the efficacy of any threat-based strategy against a nonpeer adversary. Yet the assessment of the will of the United States to utilize its capabilities and carry out any threat-based strategy is indeed a complex calculus. Thomas Schelling's notion of "the threat that leaves something to chance" is often provided less through strategic calculation by the United States than an actual uncertainty in the executive as to the most politically feasible and advantageous course. The "invitation to struggle" provided by our Constitution assures a degree of uncertainty and unpredictability in U.S. foreign policy. Further, efforts by the United States to build support for intervention through the formation of coalitions and to obtain the imprimatur of international organizations add another dimension to any effort to calculate the will behind a threat-based strategy employed against a nonpeer actor.

The dynamics of the policy formulation process within the U.S. government and perceptions of U.S. actions abroad are significantly affected when the United States faces a nonpeer adversary. In cases involving a large disparity in military power between actors, one cannot directly apply parity or near-parity-based Cold War deterrence and coercion theories designed to explain or predict U.S.-Soviet interaction or the persuasive use of force in the Cold War milieu. Further, the counterstrategies likely to be employed by nonpeer adversaries in the face of U.S. threats differ as well. Nonpeer adversaries can be expected to employ strategies designed to exploit the United States' own asymmetric nonmilitary vulnerabilities, largely deriving from the structure of our government and the nature and values of our society. These "centers of gravity" include the roles of public opinion and public diplomacy in U.S. policy making; our relatively low tolerance for casualties; concern about collateral damage associated with U.S. military action; and precariousness of coalition and alliance relations.[16]

Illustrative of the complexity of gauging the coercive will of a modern democracy was the decision by the Clinton administration to publicly rule out the use of ground troops as a coercive tool during the U.S. domestic debate preceding the Kosovo air campaign in 1999. This "signaling" was seen by many as indicating to Milosevic a potential lack of U.S. resolve and thus received wide criticism as an unneeded showing of the cards by U.S. leadership. Clearly, a detached game theorist would recommend against such a move, yet this decision was representative of the extent to which strategies for the application of coercive pressure inevitably become shaped by myriad considerations and pressures within a democratic government.[17] If the Clinton administration had not clearly suggested that a ground campaign was not seen as a viable option, that would have had a negative impact on public support for a coercive effort and thus would have impacted the level of public and congressional backing for an air campaign.

The effectiveness of any threat-based strategy is a product not only of the content and nature of the threat, but as well of the decision-making processes of the targeted actor. Any threat-based strategy must incorporate an understanding of the interests and incentives of the target's leadership structure, as well as knowledge of the processes through which decisions are made by that structure. Policy makers cannot assume that a given strategy will impact varied regimes in an identical fashion. Alexander George notes that in implementing a coercive strategy, policy makers "must replace the assumption of pure rationality with sensitivity to the psychological, cultural, and political variables that may influence the adversary's behavior when he is subjected to one or another variant of the strategy."[18]

Further, any implemented strategy will impact U.S. public opinion and therefore will affect the strategic options of U.S. leaders when dealing with future crises. This is particularly true in cases in which large-scale use of force against a significantly less capable opponent is contemplated. The current strategic position of the United States of overwhelming military superiority, coupled with the legacy of the American experience in Vietnam, which advocates either the use of overwhelming force or nonintervention, may well present U.S. policy makers with a profound quandary in future confrontations with nonpeer adversaries.[19] Presently, U.S. public opinion is not likely to support the use of military force to coerce an adversary in situations in which *vital* national interests are not at stake unless the U.S. possesses a massive advantage in military capabilities and can expect a relatively low vulnerability to casualties.[20] James Nathan captures the dilemma posed by current constraints on American military strategy imposed by domestic political considerations in his suggestion that in the future the U.S. will often be faced with "the same unhappy choice confronted during the heyday of the Cold War—the use of apocalyptic force or capitulation."[21]

Yet since Vietnam and the end of the Cold War, significant changes in technology have affected the nature of U.S. employment of military force and its effects on systems in which it is exerted or threatened. In 1990s conflicts with nonpeer adversaries, the United States demonstrated an ability to inflict substantial damage to opposing actors without incurring commensurate losses, especially through the use of air power. This asymmetry of capability when engaging in interventions is an important factor that influences the formulation of U.S. policy, as well as the dynamics of U.S. military interventions. Lawrence Freedman has noted that Cold War literature on nuclear deterrence tended to connote symmetry with stability and suggests that "the search for decisive asymmetry" during the Cold War was judged by many to be destabilizing and "too dangerous."[22] The current debate over the strategic position of the United States assumes a "decisive asymmetry," at least in the near term, and instead revolves primarily around the implications of that asymmetry for U.S. strategy and policymaking.

Reflective of this is the ongoing debate between air power and land power advocates over the ability of air power to unilaterally stabilize and control events on the ground. John Warden's notion of air occupation, the suggestion that air power can be used to effect a total domination of events on the ground, is anathema to ground power advocates who contend that air power can never substitute for ground troops when a high level of control over foreign territory is sought.[23] Efforts by the Air Force to examine the concept of strategic control, the utilization of air power assets to influence and ultimately dominate events on the ground, build upon more traditional precepts of Air Force doctrine regarding the importance of the establishment of air superiority to ground engagements.[24] Yet this idea goes considerably further, suggesting that once air superiority is established, air- and space-based assets can be used to effect control at the strategic, operational, and tactical levels, both in the air and on the ground. In advancing this concept, air power advocates have built upon the theory of sea control originally advanced by Alfred Thayer Mahan in the nineteenth century and incorporated his logic in an effort to define a new framework for the contribution of air power in regional interventions as well as larger-scale war fighting.[25]

In *Military Strategy: A General Theory of Power Control,* J.C. Wylie suggests that one of the shortcomings of classic theories of air power is an unexplored tacit assumption that control of events and people on the ground can be effected through destruction imposed from the air.[26] Yet the appeal of air power to policy makers as the first choice tool for military interventions in many of the crises of the 1990s focuses attention on precisely this dilemma. Questions remained in the minds of many analysts regarding the true contribution made by the U.S. use of air power against the Bosnian Serbs in 1995 following Operation Deliberate Force. Many view Croatian military offensives in the Krajina that were underway concurrently as the more decisive factor in Serb capitulation. Additionally, analysts are divided over air power's ability to go beyond persuasion to create a semblance of actual control over a chaotic foreign environment, and to supplant the traditional role of ground forces to establish domination over events occurring on the ground when the use of ground forces is infeasible. During the past century, several different logic patterns have emerged as strategies for the employment of air power in coercive campaigns seeking to influence decision-making processes and impact events on the ground. Though each strategy has a distinct evolution and rationale, they are frequently employed in concert, complicating efforts to determine the specific efficacy of each strategy, and the types of preconditions most conducive to coercive success.

STRATEGIES FOR AIR POWER EMPLOYMENT

Robert A. Pape is the author of *Bombing to Win,* a 1996 study of the relationship between air power and coercion. Pape has suggested that four primary strategies for the employment of coercive air power have thus far emerged in the writings of air power theorists: punishment, denial, risk, and decapitation.[27] Punishment theories focus on the direct use of air power against the civilian population of an adversary in order to break the population's morale or to foment an uprising against the opposing government and are most closely identified with the writings of the Italian artilleryman Giulio Douhet. Douhet suggested that air power would revolutionize warfare by bypassing opposing armies and taking the brutality of warfare directly to civilian population centers.[28] Punishment strategies were also advocated by Brigadier General William "Billy" Mitchell and by the first head of the Royal Air Force, General Hugh Trenchard.[29] The essence of a punishment strategy is, through a campaign of strategic bombing, to destroy "not only the *capability* of an enemy to wage war but also the enemy's *will* to fight"(emphasis added).[30] Douhet's doctrine was amended in the 1930s by officers at the U.S. Army Air Corps Tactical School (ACTS), who focused on the ability of air power to devastate industrial economies by attacking key strategic "nodes," thereby enacting a massive toll both on the industrial base of the opponent's military and on the quality of life and thus the will to fight of the opposing population.

The ACTS doctrine incorporated aspects of denial, a second strategy for air power employment. Just as a strategy of punishment mirrors the concept of a countervalue capability in classical deterrence theory, denial parallels the notion of a counterforce strategy. A denial strategy focuses on the utilization of air power to attack an opposing military, thereby reducing an enemy's war-fighting capacity. Pape characterizes denial as "smashing enemy forces, weakening them to the point where friendly ground forces can seize disputed territories without suffering unacceptable losses."[31]

A third strategic mode for the employment of air power is a graduated escalation in the use of force in order to manipulate the level of risk faced by an opponent, thus attempting to leverage an opponent's fear of future costs in order to coerce action in the present. This strategy is based largely upon the work of Thomas Schelling, who argues in *Arms and Influence* that "the power to hurt is most successful when held in reserve. It is the *threat* of damage, or of more damage to come, that can make someone yield or comply."[32]

The application by air power theorists of Schelling's perspective on the value of manipulating risk has unfortunately focused almost solely on integration with a punishment strategy. Yet punishment, by virtue of its focus on the civilian population of an adversary, would be unusable in its absolute form in most nonpeer confrontations because of domestic political constraints. Robert Pape, in the tradition of most air power theorists, focuses on the efficacy of given strategies for the employment of air power and thus largely disregards the role of domestic constraints on the feasibility and sustainability of various modes of air power strategy. Pape finds that punishment is a poor choice of coercive strategy because he believes it will fail to coerce an adversary into action. Similarly, because Pape views Shelling's risk strategy as solely a "weaker form of punishment," he therefore seems to dismiss it out of hand as a coercive strategy.[33]

A fourth strategy for the employment of air power has emerged largely from the writings of now-retired Air Force Colonel John Warden. Warden advocates utilizing air power against key leadership and command and control targets, thereby achieving "paralysis" of a target's decision-making capacity without resort to the massive force requirements of more traditional denial or punishment strategies.[34] Such a strategy, which seeks to utilize air power to cause "strategic paralysis" and to create favorable conditions for rebellion and displacement of an undesirable regime, has also been described as a "decapitation" strategy. Warden's writings focus on understanding the organizational nature of foreign governments and finding centers of gravity that will, if targeted, so inhibit decision-making processes as to functionally paralyze the government and command and control processes of the adversary state.

Warden was the prime author of Instant Thunder, the coalition bombing campaign conducted against Iraq during the 1991 Gulf War. The campaign was so named to distinguish it from Rolling Thunder, the failed effort during the Vietnam War to use gradual escalation, akin to the risk strategy advocated by Schelling, to coerce North Vietnam's communist government. Warden has written extensively on the idea of "parallel warfare," the simultaneous application of force across the complete spectrum of high priority or center of gravity targets, enabled by technological advances that free air assets from a traditional first focus on surface-to-air and air-to-air threats. Importantly, the concept of parallel warfare provides a holistic, systemic perspective on the use of force and builds on Warden's decapitation strategy to suggest a way "to achieve effective control over the set of systems relied on by an adversary for power and influence—leadership, population, essential industries, transportation and distribution, and forces."[35]

There is also a normative argument for Warden's idea of decapitation as an alternative to denial- or punishment-based strategies. Much attention was focused on the roles of Slobodan Milosevic and Bosnian Serb leaders during the 1992–95 crisis in Bosnia. The direct targeting of foreign leadership connoted by a decapitation strategy makes many uncomfortable, as assassination is proscribed by the traditional conventions governing conduct between nations. Yet, a deterrent effect of such a decapitation strategy, as employed in Iraq, is that dictatorial leaders may in the future pause before engaging in aggressive or oppressive actions if they suspect that doing so may put themselves at personal risk.

Further, Warden contests Pape's characterization of the air power strategy that he advocates as "decapitation." Pape suggests that the targeting of leadership and command and control structures is only one component of the "parallel attack" strategy that he advocates.[36] Warden argues that the ability of the U.S. military to project power deep into the territory of an opponent from the outset of a conflict allows the simultaneous targeting of all critical nodes that most directly affect the ability of the targeted system to defend against the attack, and to resist the strategic demands set forth by U.S. decision makers. Warden mentions the oft-cited improvement in the ability of modern precision air power to destroy a given designated mean point of impact (DMPI) as compared to the strategic bombers of World War II. Warden notes that a B-17 engaged in daylight bombing raids on Germany in 1943 required nine thousand bombs dropped to obtain a 90 percent probability of impacting an area approximately one-third the size of a football field. Today, a strike aircraft employing precision-guided munitions requires only one sortie and one bomb to accomplish that same mission.[37]

This example even understates the ability of modern air power assets to impact a system as compared to the air power tools of fifty years ago, as the massive advancements in target selection and location, which have played a vital role in the prosecution of U.S. air campaigns in recent years, are not represented. Modern technologies for determining target sets, finding and mapping targets for air strikes, and then evaluating the success of the strikes in a timely fashion have substantially improved the ability of military leaders and policy makers to integrate the use of air power with given strategic objectives. Any effort to target decision-making processes with air power, however, requires a fundamental understanding of not only the characteristics of the targeted actor, but of the structure and dynamics of organizations more generally, so that predictions can be made regarding the behavior of organizations subjected to the stress of a coercive air campaign.

STRUCTURE-BASED TARGETING

John Warden has suggested that the construction of states can be reflected by a model of five concentric "rings" that represent the various system components and their relative importance to the system's whole and suggests that the model can be applied to map out the vulnerabilities of any organization. He advances that "every organization has a leadership function to give it direction and help it respond to change in its external and internal environments; each has an energy conversion function to take one form of energy and convert it into a different kind of energy; each has an infrastructure to hold it together; each has a population; and each has fielded forces to protect and project the organization."[38]

Such efforts to map out the structure of an organization or system are an essential precursor to any attempt to exert influence on that system. Once that process is complete, decision makers must design a plan for leveraging their own resources against the vulnerabilities of the system, as revealed through study of its structure and character. To be successful, a plan for doing so must have coherence and consistency, with each component of the overall plan, from the micro level to macro level, contributing to its overall efficacy. In the study of the use of armed force, a four-tier hierarchy of definitional levels is often employed to specify the nature and scope of decisions and actions undertaken in the planning or conduct of warfare. This hierarchy, consisting of the grand strategic, strategic, operational, and tactical levels, is also helpful for examining the effects of other persuasive tools by policy makers used in place of or concomitant with armed force in an effort to deter or alter actions of foreign states or nonstate actors.

The grand strategic level of war describes those decisions that most fundamentally affect a state's war policy and the overall goals and direction for its actions. The next step down, the strategic level of war, is concerned with how a state's resources will be used in pursuit of grand strategic goals. The strategic level involves decisions about whether force will be used, and to what extent. In the case of strategic coercion, the strategic level is also concerned with how force or threats of force will be integrated with other persuasive methods to achieve grand strategic aims. The operational level of war is the next step in the hierarchy and concerns the methods through which specific types of military forces, or other strategic resources, are employed across the theater level in pursuit of strategic ends. Finally, the tactical level describes the character of the actual engagement of forces, providing analysis of warfare in its most fundamental state.[39] Clearly there is overlap between these levels, and individual issues may extend beyond one particular level. However, this hierarchy provides a useful analytic framework for sorting and grouping given strategies and studying the effect of strategies at each level on the dynamics of the overall system.

A given coercive strategy on one level may affect concurrent strategies at another level. Further, any coercive strategy must incorporate an understanding of existing coercive or deterrent processes at each level. For instance, the bombing of targets in Yugoslavia by NATO during the spring of 1999, an effort to coerce Slobodan Milosevic's government into removing military forces from Kosovo, had an unintended effect of opening the door for Milosevic to forcefully drive massive numbers of ethnic Albanians from the Kosovo province and to execute Kosovar citizens seen as a threat to the Yugoslav state. The coercive bombing strategy negatively interacted with the deterrent effect provided by the presence of international news media on the ground in Kosovo, thus providing an open door for Milosevic to implement an "ethnic cleansing" strategy in a relative news blackout.

Intrinsic to the five rings model that Warden advocates is the idea that an understanding of the structure of an opponent is vital to the process of constructing an effective coercive air strategy. Warden advances that his model is broadly applicable to all governments and organizations, that it depicts the "fractal" nature of the set of systems and subsystems that comprise the elements of a state, and thus the model can be broadly applied and relied upon for air campaign planning irrespective of variances in the specific structure of the targeted state. Warden advances that "the five rings approach recognizes fractal relationships which repeat themselves from the very large to the very small. In other words, each part of the system is defined by its own five rings structure right down to the level of an individual."[40]

A danger of adherence to a single specific model used to represent any system targeted by a coercive military effort is the fact that the nature and construction of systems targeted for coercion by the United States in the past has widely differed and will likely differ even more markedly in the future. Iraq, Bosnia, Serbia, and Vietnam have each been a geographic locus of past U.S. coercive efforts, and each has exhibited some of the characteristics of Warden's model. However, the understanding and use of a conceptual model is not a satisfactory substitute for specific knowledge of the state or organization that the United States seeks to influence and is only useful if a thorough knowledge of the targeted system is applied to the theoretical model.

DENIAL VERSUS PUNISHMENT

Robert A. Pape's 1996 study of the relationship between air power and coercion focuses on testing the hypothesis that denial is a more effective coercive strategy than punishment. Pape finds that denial is generally more effective than punishment, though he points out "the most effective denial strategy in any dispute depends on the strategy of the opponent."[41] Pape advances in a more recent essay that denial strategies "seek to thwart the enemy's military strategy for taking or holding its territorial objectives, compelling concessions to avoid futile expenditure of further resources."[42] One of John Warden's counters to Pape's argument is an assertion that wars and conflicts frequently center on issues other than control of territory, and therefore a strategy that focuses on the ability of an opponent to hold territory cannot always be predominant.[43] Warden argues forcefully that strategic attack is an important and viable strategy because of the ability of strategic bombers to simultaneously, or in "parallel," target all centers of gravity of an opponent. Warden suggests such an effort can "reduce the energy level of the entire system enough to reach our peace objectives."[44]

Another criticism of Pape's thesis was advanced by Barry Watts, who suggests that Pape's methodology is inappropriate for the subject that he studies, military coercion. Further, he believes that Pape does not correctly apply the theory that he advances to the case studies under consideration and thus obtains biased and invalid results. Much of Watts' discussion exists as a general caution to any social scientist seeking to create a valid predictive theory that attempts to anticipate or explain the outcome of very complex, multifaceted events. One of Watts' most important observations with respect to Pape's implementation of his theoretical model is that the various air power strategies identified in the model—punishment, denial, risk, and decapitation—have frequently been used in concert, and thus their specific efficacies are difficult to obtain in the manner that Pape suggests.[45] Pape does note that the Desert Storm air campaign, Instant Thunder, included elements of decapitation and denial, and he attributes the predominance of the campaign's effectiveness to the coalition's attacks on Iraq's fielded forces in the Kuwaiti theater of operations.[46]

Pape's efforts have proved a useful source of debate with regard to the dynamics of military coercion and the effectiveness of various methods of air power employment. Despite the criticisms of Pape's argument, his efforts to explain the utility of denial efforts in the prosecution of an air campaign are relevant to policy makers. Recent air campaigns, in Bosnia and Serbia/Kosovo, have incorporated a substantial denial component within their overall strategy, in addition to "parallel" attacks on leadership and command and control structures similar to those advocated by Warden. Yet, an additional criticism of Pape's theory is that it is based too deeply in Cold War–era paradigms regarding the use of force, and that denial strategies as he defines them become blurred in the complex conflicts that have permeated the post–Cold War era. If paramilitary forces, as opposed to conventional fielded forces, are the primary actors engaged in atrocities and predatory behaviors toward civilians, as has been the case in Bosnia and Kosovo, then a strategy centered on targeting conventional military forces in the field might fail to attrite or deter the very groups whose depredations are at the heart of the humanitarian tragedy.

Instead, a deeper exploration of the content of a denial-based logic in pursuing an air campaign might find that in contemporary conflicts, a strategy that seeks "denial," which in Pape's words is the use of "military means to prevent the target from attaining its political objectives or territorial goals," should incorporate a broader target set than solely fielded military forces, because the opponent uses a commensurately broader set of its own capabilities to pursue its objectives.[47] A decision maker who seeks to include a denial component within an air campaign should assess the particular means that any adversary uses to accomplish its political and military objectives, if applicable, and base the targeting strategy on substantial study of the structure and dynamics of the targeted organization. Pape notes, "For coercion through denial to succeed, the coercer must exploit the particular vulnerabilities of the opponent's specific strategy."[48] Further, he argues that denial strategies are generally less effective against a guerilla strategy than a conventional, or mechanized, war strategy. He attributes the success of the 1972 Linebacker raids against the North Vietnamese in convincing the North Vietnamese to accede to U.S. demands at the negotiating table to the fact that they had shifted from a guerilla war strategy, against which the Rolling Thunder raids were ineffective, to a more conventional war strategy, against which denial-based coercive efforts were more effective.[49]

Pape mentions that the dynamics of coercion, and the ability to discern the effectiveness of various strategies for the employment of air power, differ in cases when the coercer has a "monopoly or near monopoly of power." Further, he states that in situations when "the coercer can inflict unlimited harm on the target at little or no cost to itself, then the cause of coercive success is trivial and tells us little about the more important case in which the coercer does not have a monopoly of force."[50] Though the United States did not have a monopoly of force in Kosovo, U.S. forces quickly achieved air superiority and relative freedom of action within Serbian airspace. The limitation on the severity of the power projection efforts of U.S. strike assets in Kosovo was not enemy resistance, but rather self-restraint reflected by the debate within the United States and NATO over the level of public and political support for striking various target sets. Decisions made by leaders in a democracy about how military force will be used, or threatened, are among the most grave that a leader will be faced with, and any decision to use coercive strategies must be embarked on only after much consideration and reflection. In a modern democracy, *how* power is used affects the government's capacity to exercise it, both in a given crisis and in subsequent crises. In crises of the post–Cold War world, the United States and its western allies have frequently faced a dilemma: how can a democracy or an alliance of democracies exercise overwhelming power outside their borders without corrupting their self-declared normative basis for intervention? In a democracy, choices regarding national policy and military strategy are inherently normative—they are reflective of a society's values. To craft a strategic policy consistent with those values, policy makers must consider the sum total of all effects, primary and ancillary, of any use of threat-based diplomacy or actual force.

COERCIVE SYNERGY

A panoply of instruments—military, economic, and diplomatic—exists as potential tools for use by policy makers in pursuit of specified policy goals. Sanctions, both economic and diplomatic, sometimes serve as an alternative, and are often used as a precursor, to military threat-based coercive strategies. In most situations, threat-based strategies tend to be coupled with other foreign policy instruments when used in an attempt to persuade or coerce another actor. Thus, it is important to consider these other strategies and policy instruments as contextual variables when examining the impact and effectiveness of air power employment and other threat-based strategies. The Center for Strategic and International Studies undertook a useful study in October 1998 examining the relationship between economic sanctions and other "persuasive or punitive" policy instruments.[51] One finding of the study is that targeted sanctions are often more effective when coupled with either an explicit or implicit threat of military force; thus a synergistic relationship between economic and military threat-based coercive strategies is suggested.[52]

This idea of such a synergistic relationship is important, as each type of policy tool considered for use in a coercive manner, whether economic, diplomatic, or military, in addition to taking its own effects on the dynamics of a system in which it is introduced, will interact with other concurrent strategies and thus contribute to the cumulative or net outcome. In considering any given "policy package" of coercive or persuasive instruments, policy makers must therefore give thought to not only the effect that each individual strategy will have on opposing decision makers, but as well the interaction between those strategies on the broader strategic environment. Richard Haass has noted that one problem for those attempting to identify the impact of economic as opposed to other types of sanctions is the difficulty in isolating the effect of economic sanctions from the role of the myriad other international and domestic pressures that affect any decision-making process.[53] Each policy tool has varying characteristics with respect to its ability to be targeted precisely on a specified segment of the system in which it is implemented. These target sets for coercive instruments can include infrastructure, leadership, population (in practice, the primary target of economic sanctions), or the military.

Much has been made of the ability of technological advancements in the ability of air power assets to precisely impact a specific target set and thus degrade or destroy specified components of the opposing organization. The substantial reduction in circular error probable of air-delivered munitions over the past two decades allows for a much greater efficiency of effort during an air campaign, and a substantially lower threat of collateral damage when a given target is prosecuted.[54] Recent developments in weapons and delivery technologies have greatly expanded the ability of the United States to precisely target ordnance independent of environmental impediments that have hampered the application of air power in the past. Further, modern technologies for finding and mapping targets for air strikes, and then evaluating the success of air strikes in a timely fashion, have substantially improved the ability of military leaders and policy makers to integrate the use of air power with given strategic objectives.

Evolving communication technologies now enable policy makers and strategists to better assess the operational (and even the tactical) milieu from afar and to incorporate that information into real-time guidance to war fighters in the field. This has the capability to compress the feedback loop for strategic decision-making processes, necessary for any effort to establish a degree of strategic and operational control over a chaotic environment. Yet extensive feedback is also needed as to the effects of the use of air power on targeted decision makers. This requires very resilient and redundant intelligence collection sources, whose task is made more difficult by the siege mentality of an organization under attack. Our diplomatic structures should, ideally, continue to interface with targeted decision makers during the course of a conflict in order to assess their mindset and to provide opportunities for acquiescence. Avenues need to be provided to targeted decision makers to enable them to capitulate with some degree of their authority and legitimacy still intact, unless a primary goal is to end the targeted regime or organization and replace it with another. In this case, the domestic political environment and the distribution of power structures must be considered as primary factors in the development of an exit strategy and a plan for postconflict governance.

The inclusion of a diverse coalition of participants has been crucial in the maintenance of public and international support for air campaigns in Iraq, Bosnia, and Kosovo but makes operational-level military coordination more difficult because of diverse capabilities, diverse objectives, and varying preferences regarding means and strategy. In his book *Waging Modern War*, General Wesley Clark, the Supreme Commander of Allied Forces Europe during the Kosovo air campaign, notes the challenges that a coalition-based target vetting process poses for an operational commander. As an example he cites disagreements with French leadership as to whether to prosecute "politically sensitive targets" such as "top headquarters, communications and the television stations...the presidential residences and retreats (with their bunkers and communications), the electric power system, and other key targets in Belgrade."[55] Yet advantages in terms of not just added capabilities but in the increased public support and moral authority provided by the creation and maintenance of a coalition during an air campaign highlight the importance of diplomatic efforts to underpin coercive strategies, and these advantages have proved a key center of gravity in past U.S. coercive efforts. This fact has not been lost on the targets of those coercive campaigns, and strategies designed to drive a wedge between the United States and its allies have been centerpieces of countercoercive strategies employed by targets of U.S. coercion.

Organizations and states targeted by future U.S. coercive efforts can be expected to employ counterstrategies that exploit the vulnerabilities of

U.S. strategy and decision-making processes. Daniel Byman and Matthew Waxman suggest that " 'U.S.-style' coercion" is characterized by five factors: a preference for multilateralism, an intolerance for casualties, an aversion to enemy civilian suffering, a reliance on high-technology options, and a commitment to international norms. They suggest that adversaries will challenge U.S. coercion in predictable ways derived from knowledge of these patterns. They group these counter-coercive strategies under three headings: civilian suffering–based strategies, coalition-fracturing strategies, and casualty-generating strategies.[56] Admiral James Ellis, the Commander-in-Chief of NATO's Allied Forces Southern Command during the Kosovo air campaign, noted that counter-coercive strategies employed by the Serbians during Allied Force included "sporadic use of air defense assets; deceptive media campaigns; deliberately increasing the risk to NATO pilots of collateral damage; and developing political cleavages between NATO allies."[57]

Efforts to minimize collateral damage in the course of an air campaign are vital to the maintenance of public support, both domestic and international, but are as well an important source of inefficiency and strategic vulnerability. Timothy Thomas points out that millions of dollars worth of ordnance was dropped in the Adriatic "and on open countryside" because of restrictions posed by NATO rules of engagement.[58] Hyperbole aside, no air campaign is ever a "surgical" endeavor, and precision weapons that go awry because of targeting inaccuracies, aircrew error, or weapons malfunctions do not always result in precision misses. Efforts are always made to minimize the effects of collateral damage and to tailor the correct weapon to the collateral damage estimate of a given DMPI. General Clark notes that the target selection and review process during the Kosovo air campaign was "time consuming and intricate." In addition to NATO procedures, U.S. targeting procedures would assess "location, military impact, possible personnel casualties, possible collateral damages, risks if the weapon missed the target, and so forth. This analysis then had to be repeated for different types of weapons, in search of the specific type of weapon and warhead size that would destroy the target and have the least adverse impact elsewhere."[59]

The much discussed "casualty aversion" of the U.S. public and its decision makers, frequently cited as an outgrowth of failed U.S. efforts in Vietnam and more recently in Somalia, represents another tangible factor in the planning and execution of U.S. strategy in a modern air campaign. In an effort to minimize the threat to NATO aircrew, standing orders limited the minimum altitude of Allied Force aircraft to the 10,000- to 15,000-foot range.[60] Because Serbian air defenses were turned on only as needed, they represented a "constant but dormant threat," requiring that NATO aircraft stay high to maintain an altitude sanctuary from latent surface-to-air threats.[61] NATO came under some criticism for a perceived unwillingness to risk the lives of aircrew to maximize weapons accuracy, and thus collateral damage potential. Yet Anthony Cordesman accurately points out that "medium altitude flight profiles...give the pilot more overall situational awareness than low altitude flight, and the extended time over target also has advantages."[62] Further, it is important to put the perceived casualty aversion of the United States into historical context. Given the number of regimes and leaders over the course of the last thousand years of military history that have shown little to no regard for the lives of conscripted soldiers, or human lives more generally, we are fortunate to be debating the strategic ramifications of the casualty aversion of the American body politic with regard to the lives of its sons and daughters, and the operational shortcomings of collateral damage considerations. Further, public opinion polls seem to indicate that given a mission that the American people support, clear, communicative leadership, and an executable plan, Americans and their leaders would be willing to accept substantial casualties if necessary to meet the mission objectives.[63]

THE ROAD AHEAD

The United States should continue to invest in improved sensors for acquiring targets while airborne and ensuring that fragged (assigned) DMPI's coordinates are targeted correctly. This is increasingly important as air assets increasingly rely on global positioning system weapons that require no visual acquisition by the aircraft delivering the weapon. Such systems would as well be useful in prosecuting "emerging" targets. These are nonfixed targets that have the capability to move or relocate within the standard targeting cycle of an air campaign, thus frustrating preplanned efforts to assign specific coordinates to a DMPI. Such systems include improved infrared sensors, such as the advanced tactical forward-looking infrared pod, now under development. Also, improved data-link capabilities and increased joint data-link interoperability would be useful for maximizing situational awareness at all levels, strategic to tactical, and passing real-time intelligence and targeting data to aircraft in position to strike emerging targets. More flexible, higher-resolution on-board sensors would also improve the United States' ability to defend against techniques employed by Serbia and Iraq for defeating U.S. coercive air strategies, such as the fabrication of decoys, the relocation of mobile targets, and the exploitation of civilian vulnerabilities and collateral damage considerations made by NATO and the United States in the targeting process.

To support a coercive air campaign, information technology and informed analysis must efficiently distill, filter, and process information, as the massive information flows that are embedded in modern command and control processes will otherwise contribute to the "fog" of combat. Additionally, advances in communication technology have permitted unprecedented interaction between strategic decision makers and in-theater commanders at the operational and tactical levels. These advances, such as data-link of operational and tactical information to senior decision makers and frequent video teleconferencing between key players, increase the clarity and transparency of decision-making processes at each level and improve communication but can also serve to blur distinctions between responsibilities and roles at each level of command. Tactical-level actions and events seated in a politically sensitive environment may have far-reaching strategic consequences and thus often have strategic-level attention. However, as information technologies improve the ability of strategic-level decision makers to monitor tactical-level events, command and control procedures must protect the necessary degree of autonomy and decision-making authority for tactical operators to employ and fight effectively.

We must continue to develop and refine techniques for improving feedback on the effects of air attacks on decision-making processes. This must include feedback regarding the actions of military and paramilitary forces within the targeted area. U.S. capabilities for evaluating the effectiveness of air attacks must include real-time intelligence not just on effects, but on decision-making processes, and importantly via human intelligence reports, as the massive capabilities of the United States in exploiting signals intelligence are an important tool in guiding the hand of strategists during an air campaign, but they are not a technological panacea. Enhanced capabilities in this vein could potentially have helped the United States better counter Serbian atrocities against Kosovar civilians after the commencement of the air campaign.

The NATO air campaign in Kosovo clearly contributed to Milosevic's decision to capitulate to Western demands, yet other factors were critical as well, and it is important that the wrong lessons not be taken from NATO's success. Though ground troops were "ruled out" prior to the commencement of the air campaign in Kosovo, a factor in NATO's coercive success was increased discussion and support in NATO capitals over the course of the campaign for a ground operation in Kosovo in the event that air attacks and concurrent diplomatic efforts were insufficient to persuade Milosevic to agree to NATO demands.[64] Because they were not needed for coercive success, ground forces were not employed until the postconflict phase in Bosnia and Kosovo, when U.S. and NATO forces were introduced in a peacekeeping role in wake of the opposition's capitulation. Yet these cases do not decrease the imperative that air and ground assets have the capability to work effectively together in concert. The decoupling of air and ground power in Bosnia and Kosovo must not impede the ability of U.S. air and ground forces to work concurrently and synergistically in future conflicts, and realistic, joint predeployment training exercises are an important contributor to interoperability between air and ground forces.

Robert Pape argues convincingly that it is when strategies for air power employment effectively *deny* the objectives of the enemy's leadership that the likelihood of coercive success is highest. Further, rogue leaders who find themselves targeted by coercive air campaigns frequently do not value the lives of their citizens in the manner that one might intuitively hope. Coercive strategies that seek the generation of civilian hardship, whether punishment-based air attacks or economic sanctions that generate large-scale civilian suffering, must be very thoroughly scrutinized on grounds of both morality and efficacy. Punishment strategies are not preferred not only because the morality and ethics of our society argue forcefully against them, but also because their effectiveness is questionable, especially when employed against nondemocratic states for whom the suffering of the population may serve to enhance rather than detract from the strategic position of the targeted leaders.

Strategies employing denial-based logic should not be limited in scope to fielded forces and military assets, however. A detailed analysis of the means by which an opposing leadership structure seeks to accomplish its political, strategic, and personal objectives should be made in advance of air campaign planning if at all possible, and efforts to deny those objectives through air attack must be tailored to the specific incentive structure of the targeted organization and its leadership. John Warden points out the remarkable ability of modern air power assets to reach beyond the defenses of an opponent and to attack key leadership and command and control nodes from the first day of a conflict. When combined with timely, accurate intelligence, a result is that fielded forces can no longer protect leaders of predatory states or organizations from the consequences of their actions.

For much of military history, the price of campaigns fought by self-serving leaders for their own aggrandizement has been borne by the weak and the poor, and paid in tears and blood. When traveling through the Balkans, whether Serbia or Croatia, Bosnia or Kosovo, one notes that each rural village has a similar landmark—a well-tended cemetery filled with the graves of those killed in the ethnic conflicts of the past decade. If technology and military strategy can now combine to hold decision makers directly accountable for actions inimical to international security and fundamental human rights, thus forcing them to personally own up to the risks of their predations, that may prove the most effective deterrent of all.

NOTES

1. See Robert J. Art, "To What Ends Military Power?" *International Security* 4 (spring 1980): 3–35 and Robert A. Pape, *Bombing to Win: Air Power and Coercion in War* (Ithaca, N.Y.: Cornell University Press, 1996), 4.

2. See Thomas C. Schelling, *Arms and Influence* (New Haven, Conn.: Yale University Press, 1966); Alexander L. George et al., *The Limits of Coercive Diplomacy: Laos, Cuba, Vietnam* (Boston: Little, Brown, 1971); Alexander L. George, "Coercive Diplomacy: Definition and Characteristics," in *The Limits of Coercive Diplomacy* (Boulder, Colo.: Westview Press, 1994). Peter Jakobsen has noted that despite the frequent use of coercive diplomacy as a strategy for crisis management by the United States, there have been relatively few attempts to expand on the theoretical framework suggested by George and Schelling. Peter Viggo Jakobsen, *Western Use of Coercive Diplomacy after the Cold War: A Challenge for Theory and Practice* (New York: St. Martin's Press, 1998), ix.

3. Deterrence can be defined as the utilization of implicit or explicit contingent threats to persuade an actor to refrain from a specific course of action based on the actor's calculation of its rational self-interest. Coercion relies on threats to convince an adversary to undertake an action, or to abandon a course of action already embarked upon. Successful deterrence results in inaction on the part of the target state, whereas the goal of coercion is to produce a decision resulting in

positive action. Thus outcome, not method, is the source of the primary definitional distinction between the two concepts.

4. Lawrence Freedman, "Strategic Coercion," in *Strategic Coercion: Concepts and Cases,* ed. Lawrence Freedman (New York: Oxford University Press, 1998), 15.

5. Bernard Brodie, *The Absolute Weapon: Atomic Power and World Order* (New York: Harcourt Brace, 1946), 76.

6. Robert Jervis has noted that *perception* is a key variable for understanding and predicting the outcome of a deterrent strategy. Even if an actor issuing a coercive threat has both the capability and will to carry out the threat, the strategy may well fail if the targeted actor perceives the situation differently. See Robert Jervis, *Perception and Misperception in International Politics* (Princeton, N.J.: Princeton University Press, 1976), 79.

7. Gordon A. Craig and Alexander L. George, *Force and Statecraft: Diplomatic Problems of Our Time* (New York: Oxford University Press, 1990), 179.

8. See Pape, *Bombing to Win,* 4. Pape defines coercion as "efforts to change the behavior of a state by manipulating costs and benefits."

9. A contributor to the substantial confusion and semantic indeterminacy in attempting to identify and separate the concepts of coercion, compellence, coercive diplomacy, and deterrence is the fact that Thomas Schelling never specifically defined the concept of compellence and instead relied on metaphors and illustrative examples to demonstrate his concept. See Schelling, *Arms and Influence,* 69–91. Robert Pape attempts to resolve this issue by referring to coercion and compellence interchangeably. See Pape, *Bombing to Win,* 4. Schelling notes in *Arms and Influence* that J. David Singer used "persuasion" and "dissuasion" to substitute for "coercion" and "deterrence." J. David Singer, "Inter-Nation Influence: A Formal Model," *American Political Science Review* 17 (1963): 420–30. Cited in Schelling, *Arms and Influence,* 71. Stephen Cimbala also uses "military persuasion" as an umbrella concept comprising deterrence, coercion, and coercive diplomacy. See Stephen J. Cimbala, *Military Persuasion: Deterrence and Provocation in Crisis and War* (University Park, Pa.: The Pennsylvania State University Press, 1994).

10. George, "Coercive Diplomacy," 7.

11. George, "Coercive Diplomacy," 8–9.

12. Peter Viggo Jakobsen, *Western Use of Coercive Diplomacy after the Cold War: A Challenge for Theory and Practice* (New York: St. Martin's Press, 1998), 13.

13. Freedman, "Strategic Coercion," 18.

14. Jakobsen, *Western Use of Coercive Diplomacy,* 11–17. Alexander George defines *blackmail* as the offensive use of coercive threats "to persuade a victim to give up something of value without putting up resistance." George, *The Limits of Coercive Diplomacy,* 7.

15. Jakobsen, *Western Use of Coercive Diplomacy,* 17.

16. Daniel Byman and Matthew Waxman, "Defeating US Coercion," *Survival* 41, no. 2 (1989): 110–16.

17. For a discussion of the costs and benefits of transparency in the decision-making processes of a democratic government, see Kenneth A. Schultz, *Democracy and Coercive Diplomacy* (New York: Cambridge University Press, 2001).

18. George, "Theory and Practice," in *The Limits of Coercive Diplomacy*, 20.

19. The Weinberger/Powell Doctrine encapsulates much of the legacy of the American experience in Vietnam in its impact on contemporary policy making: it stipulates that force should be used only if the following tests can be satisfied: (1) Are American vital interests at stake?; (2) Are the issues so important that we will commit enough forces to win?; (3) Are the political and military objectives clearly defined?; (4) Are the forces sized to achieve the objectives?; (5) Do the American people support the objectives?; (6) Are forces to be committed only as a last resort? Taken from Richard P. Hallion, *Storm over Iraq: Air Power and the Gulf War* (Washington, D.C.: Smithsonian Institution Press, 1992), 90.

20. See Edward N. Luttwak, "A Post-Heroic Military Policy," *Foreign Affairs* 75 (July/August 1996): 33–44.

21. James Nathan, "On Coercive Statecraft: 'The New Strategy' and the American Foreign Affairs Experience," *International Relations* 7 (December 1995), 27.

22. Lawrence Freedman, *The Revolution in Strategic Affairs* (New York: Oxford University Press, 1998), 39.

23. See Marc K. Dippold, "Air Occupation: Asking the Right Questions," *Air Power Journal* (winter 1997): 69–70.

24. General Ronald R. Fogleman, "Aerospace Doctrine: More than Just a Theory" (speech at the Air Force Doctrine Seminar, Maxwell Air Force Base, Ala., 30 April 1996).

25. See Alfred Thayer Mahan, *The Influence of Seapower upon History, 1660–1783* (Boston: Little, Brown, 1897).

26. J.C. Wylie, *Military Strategy: A General Theory of Power Control*, rev. ed. (Annapolis, Md.: Naval Institute Press, 1989), 63.

27. Pape, *Bombing to Win*, 58.

28. See Giulio Douhet, *Command of the Air*, trans. Dino Ferrari (New York: Coward-McCann, 1942).

29. For instance, in *Skyways* ([Philadelphia: J.B. Lippincott, 1930], 253), William Mitchell wrote that "The advent of air power which can go to the vital centers [of gravity—a pointed reference to the work of the great theorist of war Car von Clausewitz] and entirely neutralize or destroy them has put a completely new complexion on the old system of war. It is now realized that the hostile main army in the field is a false objective and the real objectives are the vital centers. The old theory that victory meant the destruction of the hostile main army, is untenable. Armies themselves can be disregarded by air power if a rapid strike is

made against the opposing centers." Cited in Richard P. Hallion, *Storm over Iraq: Air Power and the Gulf War* (Washington, D.C.: Smithsonian Institution Press, 1992), 7.

30. Mark Clodfelter, *The Limits of Air Power: The American Bombing of North Vietnam* (New York: The Free Press, 1989), 2.

31. Robert A. Pape, "The Limits of Precision-Guided Air Power," *Security Studies* 7, no. 2 (1997/98): 97.

32. Schelling, *Arms and Influence*, 3.

33. Pape, *Bombing to Win*, 316.

34. See John A. Warden III, *The Air Campaign: Planning for Combat* (Washington, D.C.: National Defense University Press, 1988); and John A. Warden III, "Employing Air Power in the Twenty-First Century," in *The Future of Air Power in the Aftermath of the Gulf War,* ed. Richard H. Shultz Jr. and Robert L. Pfaltzgraff Jr. (Maxwell Air Force Base, Ala.: Air University Press, 1992).

35. David A. Deptula, *Firing for Effect: Change in the Nature of Warfare* (Arlington, Va.: Aerospace Education Foundation, 1995), 5.

36. John A. Warden III, "Success in Modern War: A Response to Robert Pape's *Bombing to Win*," *Security Studies* 7, no. 2 (1997/98): 175–76.

37. Ibid., 177.

38. Ibid., 175.

39. Warden, *The Air Campaign: Planning for Air Combat* (New York: toExcel, 1998), 1–2.

40. Warden, "Success in Modern War," 180.

41. Pape, *Bombing to Win,* 29–31. Pape's use of *punishment* and *denial* to describe strategies focusing on the targeting of civilian centers and military assets respectively in conventional coercive efforts is analogous to the use of the terms *countervalue* and *counterforce* by Cold War deterrence theorists to describe nuclear targeting strategies. See Stephen J. Cimbala, *Military Persuasion: Deterrence and Provocation in Crisis and War* (University Park, Pa.: The Pennsylvania State University Press, 1994), 291–92.

42. Pape, "Limits of Precision-Guided Air Power," 97.

43. Warden, "Success in Modern War," 186.

44. Ibid., 175.

45. See Barry D. Watts, "Ignoring Reality: Problems of Theory and Evidence in Security Studies," *Security Studies* 7, no. 2 (1997/98): 115–71.

46. Robert A. Pape, "The Air Force Strikes Back: A Reply to Barry Watts and John Warden," *Security Studies* 7, no. 2 (1997/98): 214.

47. Pape, *Bombing to Win,* 13.

48. Ibid., 30.

49. Ibid., 31.

50. Ibid., 49.

51. Joseph J. Collins and Gabrielle D. Bowdoin, *Beyond Unilateral Economic Sanctions: Better Alternatives for US Foreign Policy* (Washington, D.C.: The CSIS Press, 1999).

52. Ibid., xi.

53. Richard Haass, introduction to *Economic Sanctions and American Diplomacy*, ed. Richard Haass (New York: The Council on Foreign Relations, 1998), 4.

54. The CEP is the distance from a specified aimpoint within which half of the bombs dropped by a specific employment method can be expected to fall.

55. Wesley K. Clark, *Waging Modern War: Bosnia, Kosovo, and the Future of Combat* (New York: Public Affairs Press, 2001), 236.

56. Byman and Waxman, "Defeating US Coercion," 110–16.

57. Timothy L. Thomas, "Kosovo and the Current Myth of Information Superiority," *Parameters* 30, no. 1 (2000): 23.

58. Ibid., 15.

59. Clark, *Waging Modern War,* 201.

60. Anthony H. Cordesman, *The Lessons and Non-Lessons of the Air and Missile Campaign in Kosovo* (Westport, Conn.: Praeger, 2001), 99–100.

61. Thomas, "Kosovo," 23.

62. Cordesman, *Lessons and Non-Lessons,* 99.

63. Ibid., 100.

64. Ibid., 88.

Chapter 3

COALITION WARFARE: THE COMMANDER'S ROLE

Derek S. Reveron

The most aggressive military campaign of the Clinton Administration began on March 24, 1999, when Operation Allied Force commenced. The attack came after several months of negotiation that culminated in failure at Rambouillet. With its second foray into the Balkans, NATO launched an impressive air campaign over seventy-eight days. During the air campaign, NATO aircraft flew approximately 38,000 combat missions, 23,000 of which were strike missions. Throughout the air campaign, NATO's commander, General Wesley K. Clark discovered the difficulties of coalition warfare. To maintain the alliance, General Clark pursued a deliberate strategy that not only pushed the limits of NATO members at the strategic level, but also involved NATO diplomats in tactical decisions.

The importance of maintaining the alliance in the face of Serb resolve was more critical when Milosevic did not sue for peace as expected after three days of attack. When initial strikes did not force Milosevic to quit, General Clark used his diplomatic skills to build support for sustained operations. Clark said, "I talked to everybody. I talked to diplomats, NATO political leaders, national political leaders, and national chiefs of defense. There was a constant round of telephone calls, pushing and shoving and bargaining and cajoling, trying to raise the threshold for NATO attacks."[1] Without General Clark's diplomacy, it is unlikely that NATO could have sustained seventy-eight days of support for air operations. Thus, Clark fought two wars: an offensive one against Milosevic and a defensive war against NATO critics.

A NATO OPERATION

Since the mid-1990s, NATO claimed responsibility for stability in the Balkans. The United States and European countries deliberately avoided the United Nations to get international approval for an air campaign in the Balkans.[2] When confronted with the perception of intervening in a Yugoslav civil war, the United States and its allies feared that Russia or China would veto any intervention designed to stop fighting in Kosovo.[3] To ensure that the United States would be unopposed, it pursued an anti-Milosevic policy through the NATO alliance.[4] Leading this effort was NATO's commander, the American General Wesley Clark, the Supreme Allied Commander of Europe (SACEUR).

Though SACEUR was mindful of potential Russian or Chinese interference, General Clark also did not want to repeat the very bruising experience of the United Nations Protection Force (UNPROFOR) in Bosnia during the 1990s.[5] According to Clark, "the OSCE [Organization for Security and Cooperation in Europe] mission looked like it was going to place unarmed people at risk, and in addition, it was going to place people on the ground whose very presence would enable them to be taken hostage and, therefore, checkmate the air threat."[6] With a NATO force under his command, there would be no opportunity for the Serbs to repeat actions against UNPROFOR by taking peace monitors hostage again. Further, the predilection for armed intervention was based on the belief that Milosevic understood only the use of force. A NATO military official said, "In the Balkans, they only respect you when you go in with all guns blazing."[7]

The assumption that blazing guns were necessary in the Balkans was based on previous experiences with Milosevic. Clark credited the threat of force in 1998 to compelling Milosevic to accept an OSCE force in Kosovo. Threatened force was primary because many in NATO believed that its attacks in 1995 compelled Milosevic and Bosnian Serb leaders to the bargaining table in Dayton. Staged in late summer 1995, NATO undertook Operation Deliberate Force with about 750 strike missions against fifty-five preselected targets in Bosnian Serb territory. After ten days of attack, Serb leaders agreed to negotiate.

The analysis of NATO's effectiveness is wrongheaded, however. Two months prior to the air strikes, the Croat and Bosnian armies started to overturn Serb territorial gains. At the end of July, the Serbs controlled 70 percent of Bosnia's territory. By September 22, Serb gains were reduced to 49 percent. At the time, Ambassador Holbrooke recognized the necessity of reversing territorial gains. Holbrooke told Croatian President Tudjman that territorial gains "had great value to the negotiations... and it would be much easier to retain at the table what was won on the battlefield."[8] In spite of the ground gains being important to the negotiations, it was the air campaign that was credited with convincing Milosevic to give up. Admiral Leighton "Snuffy" Smith, NATO's southern commander in Naples, set the stage for future dealings with Serbia: "What we are trying to do is not defeat somebody on a military basis but trying to compel certain standards of behavior. And I think we can do that with air power."[9]

In 1999, the goal again was to compel certain standards of behavior. Secretary of State Madeleine Albright declared one month before the air campaign commenced, "The primary obstacle to peace remains Slobodan Milosevic."[10] With the threat of 1.8 million Kosovar Albanian refugees, Clark thought NATO's threat would deter Milosevic. As NATO's military commander, Clark had the ability to solicit air contributions from NATO's nineteen members. With such an arsenal at his disposal, Clark exclusively relied on air power again to get Milosevic to acquiesce to NATO's demands for Kosovo. When the threat did not result in Serb concessions, Clark was faced with the task of planning and managing an air campaign to coerce Milosevic to accept NATO's demands. Since Milosevic was considered to be Serbia's center of gravity, the attack became personal. Targets moved from integrated air defense sites to Milosevic-owned assets and residences.

The importance of punishing Milosevic through air power and making Allied Force a personal attack against Milosevic is evidenced in the language of the press conferences. French President Jacques Chirac said in the second week of the air campaign, "The horror wanted and organized by Milosevic goes beyond imagining."[11] British Secretary of Defense George Robertson emphasized that the war was aimed at Milosevic: "We have made it very clear that hollow half measures will not stop the damage we are doing to his killing machine."[12] The personal nature of the air attack was stressed when General Clark briefed the operation, our fight is "with the brutal leaders [of Serbia]"[13] and "there is no sanctuary for them, their military forces, their command and control elements, as this campaign continues because they're part of the mechanism of the Serb military and security forces and they're oppression."[14]

RELUCTANT ALLIES

Allied Force illustrates the difficulties of coalition warfare. Not only were there capability differences among air forces, but also there were reluctant partners in NATO. Strobe Talbott, the American Deputy Secretary of State at the time, said "there would have been increasing difficulty within the alliance in preserving the solidarity and the resolve of the alliance" had the Serbian leader not given in on June 3, 1999.[15] Throughout the conflict, several allies reluctantly supported the operation, but their concern for the humanitarian situation in Kosovo prevented them from stopping the operation. To allay concerns and shore up support, General Clark was instrumental to maintaining the alliance. He did this not by silencing opposition, but by encouraging skeptics to pursue their own national agendas.

Given Greece's strategic position only 100 miles away from Kosovo, the use of Greek ports and airspace was important to the operation. However, Greek public opinion was overwhelmingly opposed to NATO's actions against Serbia. Over 90 percent of Greeks opposed NATO air strikes because of a common history of fighting the Ottomans and a common religion in orthodoxy. Publicly, the government supported NATO's actions and did not oppose the operation. Underlying its tacit approval against its public's wishes was the government's goal of being a part of Europe. Greek Prime Minister Costas Simitis said, "We are both a Balkan country and a member nation of the European Union.... We have tried for years to participate in European unification. We are only a breath away from achieving it. We have no right to endanger this."[16] In other words, Greece was fundamentally against the operation, but it would not use its power within NATO to end it.

Outside of NATO channels, Greek support of the operation was limited to providing access to an airfield on Crete and facilities at Thessaloníki. Greece opposed Serb military operations in Kosovo, but it did not militarily participate in Allied Force and even sent medical supplies to Serbia. To appease public opinion, the Greek government rescinded its offer to accept 10,000 refugees and reinforced its border to prevent illegal entry into the country. After the first day of the air attacks, Greek diplomats engaged in shuttle diplomacy between Belgrade and other European capitals that culminated in a Serb cease-fire offer in observance of Orthodox Easter. Greek Prime Minister Costas Simitis called the proposal "a first step" for a political dialogue with Yugoslavia, but NATO diplomats rejected the offer as a ploy, and the air campaign continued. General Clark was adamant that a bombing pause without Serb commitment to a NATO presence in Kosovo would undermine negotiations. Clark reasoned that once the bombing stopped, it would be very difficult to start it again.

Like Greece, Germany opposed Serb actions in Kosovo, but Germany's Nazi past compelled it to participate in the NATO attacks. German

Defense Minister Scharping summarized German support of the NATO operation: "It [Serb action] is a systematic extermination, which reminds in a terrible way of what happened and was carried [out] under Germany's name... at the start of World War II and during that entire war in places like Poland."[17] Because Milosevic was likened to Hitler, and Serb operations were likened to Nazi genocide, it was easy for Germany to oppose Serb actions. However, pacifist elements in the German government objected to the use of force. German Foreign Minister Joschke Fischer, senior member of the Green Party, sought a peaceful resolution throughout the air war. After eight weeks of bombing when the United Kingdom was discussing ground force options, Germany objected. German Chancellor Schroeder said, "At this point when the NATO strategy is yielding fruits, I do not think it would be wise to change the strategy one way or another—that is to say, either by sending ground troops or by bilateral cease-fires."[18]

While German and Greek diplomats worked to find a peaceful solution, pressure also came from Italy. Given its strategic location and heavy presence of American military resources, Italy was important to the air campaign's logistics. Roughly 90 percent of NATO sorties took off from Italian territory. Despite its important role, Italian support for the operation was mixed. In May, the Italian Parliament passed a resolution calling for the government to work to suspend NATO's bombing and Prime Minister D'Alema supported a cease-fire to allow negotiations within the U.N. Security Council. Again, NATO's commander placed the requirement that a cease-fire be initiated after a comprehensive agreement was signed and Serb forces withdrew from Kosovo.

The enthusiasm of NATO's newest member, Hungary, quickly diminished when the air campaign commenced. Concerned about the 300,000 ethnic Hungarians in the Vojvodina, a northern province in Serbia, Hungarian diplomats requested that the region be spared from NATO attack.[19] Foreign Minister Janos Martonyi reminded NATO that his country has been placed in the agonizing position of going to war against fellow Hungarians.[20] In spite of these concerns, Hungary supported the operation in a limited way. As the only NATO member bordering Yugoslavia, Hungary was critical for air flight clearance and provided basing in the campaign's second month for twenty-four American attack aircraft in Taszar and refueling aircraft in Budapest. However, Hungary declared early in the air campaign that it would not allow its territory to be used for a land invasion of Serbia.

With reluctant allies, the operation was always threatened with termination. The British led the diplomatic effort to shore up the alliance, and

Prime Minister Tony Blair likened critics of NATO's strategy to those who appeased Hitler in the 1930s. General Clark summed up the difficulties he faced:

> There was a lot of concern among some in the alliance that we might be forced to accept a bombing pause. So one of my responsibilities was to argue against the bombing pause. It would have given the Serbs a chance to recover their defense system. It would have given them a chance to continue the ethnic cleansing campaign on the ground. And it would have made Western political leaders and NATO appear as though we didn't really have a strategy and a program to move ahead. And we couldn't afford any of that—we had to move forward in this campaign. And that position prevailed.[21]

General Clark not only faced a lack of enthusiasm from America's NATO allies, but from the United States itself. Very early in the conflict, President Clinton ruled out the use of ground forces. According to the U.K. House of Commons Defence Committee *Lessons of Kosovo* report, "public pronouncements made throughout 1998 and well into 1999" discounting "a forced entry ground option as part of [NATO's] military strategy, were in military terms a serious error of judgment."[22] President Clinton and other Alliance leaders indirectly told Milosevic that his way out was to endure aerial bombardment; because of this, Milosevic hoped NATO would make enough mistakes to undermine diplomatic support for the air campaign. Clark's task was to ensure no mistakes were made.

CASUALTY-FREE WARFARE

Faced with the diplomatic prerequisite of risk-free warfare, General Clark prevented alliance decay by reducing the possibility of collateral damage and civilian casualties. According to Clark, given the memories of World War II bombings, "We had to convince them [Europeans] of the validity of the targets, the accuracy of the delivery systems, the skill and courage of the airmen, and their ability to deliver weapons with pinpoint accuracy."[23] Targets were studied to determine the effects on nearby civilian facilities. If the risk was too great for collateral damage, the target was avoided or was attacked with a very precise weapon. Lord George Robertson stated, "A balance had to be struck between the risks taken, and the likely results."[24] Acting according to this principle, attacks were explicitly timed to avoid the risk of casualties. The result, in some critics' eyes, was the destruction of empty buildings.

Targeting became more conservative after mistakes. Following an attack on a rail bridge that resulted in the loss of a passenger train, air campaign commander General Short changed the tactics against bridges. He later testified to the Senate Armed Services Committee, "The guidance for attacking bridges in the future was: You will no longer attack bridges in daylight; you will no longer attack bridges on weekends or market days or holidays. In fact, you will only attack bridges between 10 o'clock at night and 4 o'clock in the morning."[25]

In addition to the fixed target strikes, the bulk of NATO's effort against tactical targets was aimed at fielded forces, heavy weapons, and military vehicles and formations in Kosovo and southern Serbia. Many of these targets were highly mobile and hard to locate, especially during the poor weather of the early phase of the campaign. Strikes were also complicated by the Serb use of civilian homes and buildings to hide weapons and vehicles, the intermixing of military vehicles with civilian convoys, and sometimes, the use of human shields. In this way, NATO's concern to avoid civilian casualties was exploited by the Serbs. In mid-April, NATO pilots misidentified a refugee column as a Serb army column, and this resulted in the deaths of about seventy-five refugees. After that incident, pilots exercised greater caution before releasing their weapons.

To reduce the chance for mistakes, there was a very high use of precision-guided munitions (8,500 out of 25,000) compared to 7 percent during Desert Storm.[26] Former Secretary of Defense William Cohen called it "the most precise application of air power in history."[27] Cruise missiles were used extensively during the first few days of the conflict. Tomahawk was used against almost 20 percent of all the targets attacked, including almost half the headquarters facilities, almost half of the electrical and power facilities, and one-quarter of the petroleum and oil facilities.[28] Finally, the B-2 made its wartime debut with the delivery of the Joint Direct Attack Munition (JDAM), which was very effective for two reasons. First, the B-2 stealth bomber delivered it, which reduced the likelihood of aircraft casualty. Second, the all-weather, global positioning system–guided weapon was extremely accurate against fixed targets. A total of forty-five B-2 sorties delivered 656 weapons. JDAM was so preferred that it was expended at its production rate. As a testament to accuracy, of the 23,000 weapons dropped, there were fewer than twenty collateral damage incidents. Human Rights Watch estimated that there were ninety incidents involving civilian deaths, in which between 488 and 527 civilians died.[29] NATO Secretary General Lord George Robertson summarized the campaign goal: "The concern to avoid unintentional damage was a principal constraining factor throughout."[30] With the precision of modern weapons, NATO limited its potential for mistakes and did not give Milosevic material for his propaganda campaign in Serbia.

For General Clark to continue to prosecute the campaign, he not only limited collateral damage but also reduced the likelihood of NATO casualties. Aircrew flew under strict rules of engagement that required aviators to have visual contact with the target. To avoid anti-aircraft artillery and shoulder-fired surface-to-air missiles (manpads), aircraft flew above 15,000 feet. The trade-off of flying at higher altitudes to mitigate risk made weather conditions such as cloud layers and visibility more of a factor in daily execution. Further, since General Clark could control public relations in the event of American casualties, U.S. aircraft flew the difficult missions. Throughout the air campaign, U.S. aircraft dropped 70 percent of the munitions.[31] Lord Robertson, British Defense Minister at the time, expressed U.S. dominance in a negative way: "There were some [NATO] countries that felt embarrassed during Operation Allied Force because they could not make the contribution they wanted." For General Clark, it was important to not risk NATO aircraft and NATO public opinion. Not able to satisfy the United States either, General Clark faced criticism from American diplomats who were unhappy with the United States assuming most of the risk. Deputy Secretary of State Strobe Talbot said, "Never again should the USA have to fly the lion's share of the risky missions in a NATO operation."[32]

NATO strike packages flew with high levels of support with a combination of active support jamming, High-Speed Anti-Radiation Missiles (HARMs), and also a variety of precision-guided munitions to destroy key elements of the Yugoslav air defense system. Electronic warfare aircraft suppressed the Serbian air defense system. Without use of its associated radar, the Serbs relied on unguided, ballistic surface-to-air missile launches. The average aircrew participating in Operation Allied Force experienced a missile-launch rate three times higher than encountered by the average coalition aircrew during Operation Desert Storm, yet were six times less likely to be shot down.[33] Of the over 700 missiles launched, only two NATO aircraft were hit.[34] Finally, early warning aircraft maintained constant watch of Serb airspace to prevent Serb aircraft from fly-

COALITION WARFARE: THE COMMANDER'S ROLE 59

ing to engage NATO planes. When Serb aircraft did attempt to fly, continuous combat air patrol by fighters resulted in six Serb fighters being destroyed after take-off. The goal for SACEUR was to keep the casualties low in order to keep the coalition together, but also to keep it an allied operation.

AN ALLIED OPERATION

As the name of the operation implies, General Clark made Allied Force an allied operation. Thirteen allies contributed 327 aircraft and flew 39 percent (15,000) of the 38,000 missions. After the United States, France was the largest allied contributor with eighty-seven aircraft, while Portugal contributed the fewest with only three aircraft.[35] Greece, Luxembourg, Iceland, Poland, Hungary, and the Czech Republic did not contribute aircraft. However, Greece and Hungary provided territorial overflight access and basing rights, which were critical to mission success. Without European airspace and basing and support facilities, the operation would have been difficult to conduct. In total, twenty-four bases in Europe were used.[36]

Since there were reluctant allies, General Clark involved ambassadors in the details of the operation. These meetings took place formally in NATO facilities, over the telephone, and even at his chateau. The North Atlantic Council reserved the right to approve "controversial" targets on an individual basis. "NATO used this mechanism to ensure that member nations were fully cognizant of particularly sensitive military operations, and thereby, to help sustain the unity of the alliance."[37] The reviews ensured that targets complied with international law, were militarily justified, and minimized the risk to civilian lives and property. Though criticized for allowing diplomats to participate in the target approval process, it is that fact that enabled the operation to succeed. According to Clark, "No single target, no set of targets, and no bombing series was more important than maintaining the consensus of NATO."[38] In order to maintain the alliance, SACEUR went to great lengths to include reluctant members in the decision-making process.

THREATS TO THE OPERATION

As previously discussed, the air operation was threatened by the potential for collateral damage and casualties. In addition to NATO actions that could undermine the air campaign, Serb actions presented a challenge to alliance stability. By controlling refugee flows and standing resolute against NATO, the Serbs forced NATO into a war of patience. Though NATO outlasted Serbia, the air campaign struggled from the beginning to achieve its objective, in the words of General Clark, "going after the forces inside Kosovo and around Kosovo to destroy these forces, to isolate them, to interdict them and to prevent a continuation of their campaign or its intensification." NATO's Secretary-General stated the objectives of the air strikes: "To prevent more human suffering, more repression, more violence against the civilian population of Kosovo...[and] to prevent instability spreading in the region."[39] If NATO could not prevent a humanitarian crisis in Kosovo, then it severely undermined the rationale for the use of force. This is true no matter how many times Clark declared "we're winning, he's [Milosevic] losing, and he knows it," or reminded the press that air campaigns can do little to stop a paramilitary campaign.

Throughout the first two months of the air campaign the Serb military (VJ) and interior police (MUP) maintained the ability to expel refugees. This fact undermined President Clinton's statement on the operation's first night that the purpose of NATO's actions was "to halt an even bloodier offensive against innocent civilians."[40] NATO's bombing efforts had little effect on the ground situation, as evidenced by the refugee flows. Five days into the air campaign, approximately half of Kosovo's 1.6 million Albanian population were internally displaced, with nearly 70,000 expelled to Albania. Refugee flows continued to spike throughout the air campaign, particularly during April 15–20 and again April 30 to May 13. Even after seven weeks of bombing, NATO could not prevent the creation of refugees. In mid-April, over 620,000 Kosovar Albanians became refugees, and the number increased in mid-May to over 800,000.[41] When the air campaign ended, nearly all of the 1.6 million Kosovar Albanians were internally displaced or refugees, 600 settlements were destroyed, and the Serbs caused approximately $1.3 billion in damage.

In addition to not limiting the refugee flow, NATO did not limit the activities of the VJ and the MUP. Designed by Tito to fight a partisan-style hit-and-run war, both forces maintained tactical effectiveness to fight the Kosovo Liberation Army (UCK).[42] When NATO did attempt to target Serb-fielded forces, weather and terrain provided concealment. With NATO intentions publicly known, military barracks were deserted, and soldiers dispersed into the countryside before the bombing commenced. Heavy weapons were concealed in forests, caves, and civilian areas.

Balkan weather provided a protective shield. Cloud cover exceeded 50 percent nearly 70 percent of the time, which resulted in unimpeded strike operations on only twenty-four of seventy-eight days of operations.[43] Further, the Serbs used weather to operate and relocate their forces. The strategy degraded NATO's ability to conduct air strikes and ensured the survival of Serb forces.

During the year preceding NATO's operation and throughout the seventy-eight days, the Serbs displayed tremendous resolve. The Serbs portrayed their enemy, the UCK, as terrorists who hid among civilians and used ambush as their main tactic. This tactic appealed to Russians who faced their own problem with Chechen separatists. Once NATO brought the conflict to Serbia's capital, defiance was exhibited on Belgrade's bridges where civilians danced throughout the night with targets pinned to their shirts.

Internationally, the Serbs portrayed themselves as victims subjected to an unprovoked NATO attack. With the exception of ballistic surface-to-air missile launches, Serbia avoided direct attack on NATO forces in Albania and Macedonia, and it kept its naval forces in Montenegrin ports. Exemplifying frustration with NATO, Serbia filed suit in the International Court of Justice. On April 29, 1999, Yugoslavia brought proceedings before the Court against Belgium and nine other NATO countries to redress a "violation of the obligation not to use force."[44] The claims were based upon the United Nations Charter and several international legal conventions, including the 1949 Geneva Convention, its 1977 Additional Protocol 1, and the Genocide Convention. Yugoslavia requested the Court's ruling on the following provisional measure: "The NATO countries shall cease immediately its acts of use of force and shall refrain from any act of threat or use of force against the Federal Republic of Yugoslavia."[45] On June 2, the Court stated that it was "profoundly concerned with the use of force in Yugoslavia," which "under the present circumstances...raises very serious issues of international law," but the Court declared that it did not have jurisdiction.[46]

Finally, NATO faced the constant threat of Russian diplomatic or military intervention. As a member of the Contact Group, Russia was committed to a peaceful solution to the problem of Kosovo. Using its influence, Russia got the Serbs to accept the political agreement at Rambouillet but could not get Milosevic to accept a NATO implementation force because of Serbia's sovereignty claim. The sovereignty claim was exacerbated by appendix b, paragraph 8 of the Rambouillet Agreement that states:

> NATO personnel shall enjoy, together with their vehicles, vessels, aircraft, and equipment, free and unrestricted passage and unimpeded access throughout the FRY including associated airspace and territorial waters. This shall include, but not be limited to, the right of bivouac, maneuver, billet, and utilization of any areas or facilities as required for support, training, and operations.[47]

That last clause subjects all of Yugoslavia to NATO. Milosevic feared that NATO would use its Rambouillet power to pursue war criminals and occupy other territories such as the Hungarian-majority province of Vojvodina.

By inserting that significant clause deep in the appendix, NATO's negotiators secured the legal power to enforce Yugoslav compliance. Reflecting a year after the conflict, Secretary-General Robertson stated, "President Milosevic had repeatedly failed to honour previous agreements and...an international security presence was essential to guarantee that the Accords would be honoured. Also, without such a presence, the Kosovar Albanian side would not have given their agreement."[48] But for the Serbs, that clause made the agreement impossible to ratify.

Once the bombing started, the fear of Russian intervention was omnipresent. Russia immediately suspended cooperation with NATO. Russia forced NATO to close its office in Moscow and expelled all NATO personnel from Russia. Russia suspended the NATO-Russia Permanent Joint Council and withdrew its personnel from NATO headquarters. One week after the bombing commenced, Russia deployed an intelligence-gathering ship to the Adriatic. As the second week of bombing commenced, Russia filed a transit request with Turkey for eight warships. However, Russia did not deploy the warships to the Adriatic, but maintained intelligence-gathering ships to monitor NATO aircraft flight activity. Diplomatically, relations between Russia and the United States were strained. President Yeltsin warned of a new cold war, and a plane carrying Russian Prime Minister Yevgeny Primakov to Washington turned back midflight in protest of NATO's actions.

The fear of Russian military intervention is best exemplified after peace was reached between NATO and Serbia. After the bombing stopped, a Russian battalion stationed in Bosnia undertook a 14-hour drive to Kosovo on June 11. Knowing that the Russian paratroopers intended to seize the Pristina airfield, General Clark ordered his British ground commander to take the airfield first.[49] NATO planned to use the airfield as the Kosovo Force headquarters, and control of the runway would be useful to quickly deploy NATO forces. However, Sir General Michael Jackson refused the order and told his boss, "I'm not going to start the Third World War for you."[50]

CINC DIPLOMACY

In many ways, Operation Allied Force was a test of NATO's new strategic concept. NATO has proven its commitment to military operations other than war through its actions in the Balkans.[51] With the collapse of the Soviet Union, many questioned the relevancy of the alliance. However, with the Red Army threat gone, the advocates of the alliance emphasized the political importance of NATO. In essence, NATO shifted its reading of the North Atlantic Treaty from Article V, which stresses unity in face of attack, to Article IV, which stresses the cooperative aspects of NATO.[52] At the center of the emphasis of NATO as a proactive political and military alliance was the Supreme Allied Commander.

Dual-hatted as Supreme Allied Commander and Commander in Chief Europe, (CINC EUR) General Clark's actions epitomize a recently recognized phenomenon called "Commander in Chief diplomacy." In terms of Kosovo, Clark was an active diplomat representing the interests of all nineteen members of NATO. An important tool given to Clark was NATO approval for air strikes in the fall of 1998. General Clark used this threat of force to help negotiate an agreement that created the OSCE Kosovo Verification Mission on October 16, 1998. Called the "Clark-Naumann agreement," it required a partial withdrawal of Serbian forces from Kosovo, limited the introduction of additional forces and equipment into the area, and deployed unarmed OSCE verifiers to Kosovo.

After the OSCE Chief of Mission William Walker discovered the deaths of forty-five Kosovar Albanians in the village of Racak, murdered in January 1999, General Clark resumed negotiations with Belgrade. Acting on behalf of the North Atlantic Council, Clark went to Belgrade to speak directly with Milosevic. The negotiations led to the talks at Rambouillet, in which Contact Group diplomats attempted to reconcile the two sides. However, the proposed Rambouillet Accords required a NATO force to monitor Kosovo; thus General Clark as NATO's commander played a critical role in the agreement's details. When the talks failed to produce an agreement, Clark was at the center of operations as NATO's commander.

Since Clark was the military commander of NATO's nineteen members, he used his position as SACEUR to build and maintain consensus for air strikes. Including reluctant allies in military planning and appealing to countries' national goals, such as British and French European ambitions, Clark not only led NATO but also the United States into battle over Kosovo. This is important because the U.S. position was that the European allies must bear a greater share of the burden in dealing with Kosovo. However, Clark used his position as NATO commander to lead the United States within NATO and used his diplomacy to build operational support among all NATO members.

As the spokesperson of the air campaign, General Clark used his position to keep international attention on the Kosovo crisis. At times, Clark speculated about genocide or was optimistic about the effectiveness of the bombing. The genocide link was likely based on experience of the 1992–95 war in Bosnia and the tragedy of Srebrenica. The preconceived notion that the Serbs were killers was reinforced by tales from refugees forced out of their villages at gunpoint. Clark and others compared Milosevic to Hitler. When the air campaign started, British Defense Secretary George Robertson said, "We are confronting a regime which is intent on genocide."[53] In mid-April, after three weeks of an ineffective air campaign, reports of 100,000 missing Kosovar Albanian men solidified the Alliance. Reported missing by the United Nations High Commissioner for Refugees, they were assumed by many to be victims of genocide. American Secretary of Defense Cohen confirmed this fear a month later when he said, "We've now seen about 100,000 military-aged men missing. They may have been murdered."[54] Despite the qualifier of "may," Cohen's statement came at an important time when the alliance started to seriously consider a ground war. Based primarily on rumors and fears, the report of 100,000 missing was false. The truth was that the men were either hiding in Kosovo or volunteered for the UCK to battle Serb forces. As of May 2003, approximately 3,000 bodies had been discovered.

Responsible for the air campaign's effectiveness, Clark overestimated the bombing success. One month after the operation ended, NATO could confirm only 60 percent of its original effectiveness estimate.[55] In the last days of the air campaign Clark insisted, "Strategically NATO's air campaign continues to focus on disrupting Milosevic's ability to direct and sustain the conflict in Kosovo—a conflict that violates all international norms. We are weakening his hold on the power base that enables him to conduct his ethnic cleansing."[56] However, when General Clark made this statement on June 2, 50 percent of the Kosovar Albanians had already been expelled, while the other 50 percent were internally displaced.

CONCLUSION

In spite of its weaknesses, history will judge Operation Allied Force successful. After seventy-eight days of air attack, Milosevic conceded to the Contact Group's demands, and a NATO force deployed to Kosovo. In the first four years of Kosovo's reconstruction, not one NATO soldier was killed. A solid agreement between NATO and Belgrade ensured that Kosovo would not become a quagmire. However, why Milosevic surrendered is not easy to say. To be sure, the humanitarian crisis was underway before NATO's attack commenced, but the scale of the crisis exploded after the first bomb fell. At the end of the seventy-eight-day campaign nearly all of the 1.6 million Kosovar Albanians were refugees or internally displaced persons. A humanitarian tragedy occurred irrespective of NATO's actions, but the alliance remained committed to the operation and began planning for a ground offensive.

The operation's commander, General Clark, constantly shored the commitment to the air campaign. By avoiding casualties and collateral damage, NATO implemented a near flawless air campaign. Further, by including potential critics in the tactical decision-making process, Clark co-opted reluctant allies to allow the operation to continue. With no chance of the alliance collapsing, Milosevic was forced to submit. This lesson Clark teaches us about coalition warfare should not be lost; coalitions are vulnerable to splintering, but once united, coalitions are invincible.

NOTES

1. General Wesley K. Clark (Ret.), interview, "War in Europe," *Frontline*, Public Broadcasting System, (http://www.pbs.org/wgbh/pages/frontline/shows/kosovo/interviews/clark.html), accessed 1 November 2001.
2. On a symbolic level, the United Nations Security Council (UNSC) passed Resolution 1160 in March 1998 "condemning the use of excessive force by Serbian police forces against civilians and peaceful demonstrators in Kosovo" and imposed an arms embargo on Yugoslavia. Six months later, the UNSC passed Resolution 1199 (1998), which underscored the international importance of Kosovo and stated that "the deterioration of the situation in Kosovo, Federal

Republic of Yugoslavia, constitutes a threat to peace and security in the region." The Security Council demanded that all parties cease hostilities and that "the security forces used for civilian repression" be withdrawn. UNSC, *On the Letters from the United Kingdom (5/1998/223) and the United States (5/1998/272)*, Resolution 1160, 31 March 1998, (http://www.un.org/docs/scres/1998/scres98.htm), accessed 8 July 2003; UNSC, *On the Situation in Kosovo*, Resolution 1199, 23 September 1998, (http://www.un.org/docs/scres/1998/scres98.htm), accessed 8 July 2003.

3. Logically, Russia would object to air strikes against Serbia for two basic reasons. First, historically, Russia views itself as the protector of all Slavs. Second, if Russia sanctions U.N. armed intervention in Yugoslavia, Russia might open itself to U.N. armed intervention in Chechnya. Chinese objections were based on its respect for territorial sovereignty and the fear that its own policies in Tibet or plans for Taiwan might be subject to international approval. The Russian representative reminded the Council that it "alone should decide the means to maintain or restore international security," and that NATO's action would set a dangerous precedent. He warned that "a virus of a unilateral approach would spread," and that those who had initiated the military venture "bore complete responsibility for its consequences." China's representative called the Kosovo question an internal matter of the FRY, arguing that the NATO action "amounted to a blatant violation of the United Nations Charter as well as the accepted norms in international law." See "NATO Action against Serbian Military Targets Prompts Divergent Views as Security Council Holds Urgent Meeting on Situation in Kosovo," (U.N. Press Release SC/6657, 24 March 1999).

4. While there was no specific authorization for NATO or any other military organization to use force in Kosovo except to protect unarmed monitors (OSCE force) in an emergency, the United States, along with other NATO supporters, asserted that under UNSC Resolution 1203, NATO had the necessary authority to use force to protect civilians in Kosovo and to enforce the cease-fire agreement.

5. From 1992 to 1995, UNPROFOR's mandate was to support the United Nations High Commissioner for Refugees to deliver humanitarian relief throughout Bosnia and Herzegovina, to protect convoys of released civilian detainees, to monitor the "no-fly" zone, and to monitor the United Nations "safe areas" established by the Security Council around five Bosnian towns and the city of Sarajevo. Because the U.N. force had limited rules of engagement and was in the midst of a civil war, it suffered 211 casualties, incurred hundreds of European prisoners used as "human shields," and did not prevent atrocities in Sarajevo, Gorazde, Bihac, and Srebrenica. In Srebrenica alone, over 2,500 bodies were found with thousands still missing. By opening the Sarajevo airport under U.N. control, General MacKenzie, UNPROFOR Chief of Staff, "thought that would cool tempers and create a situation for a negotiated ceasefire." However, the other elements of the agreement were never implemented and opening the airport "didn't change the

situation at all. In fact, if anything, it started to feed the fighters." Quoted in Sharon Hobson, "The Jane's Interview," *Jane's Defence Weekly,* 19 September 1992. For U.N. casualty totals, see http://www.un.org/Depts/dpko/fatalities/totals.htm, accessed 15 October 2001. For more on the atrocities in Srebrenica, see *Report of the Secretary General Pursuant to General Assembly Resolution 53/35: The Fall of Srebrenica,* (http://www.un.org/peace/srebrenica.pdf).

6. Clark, interview, "War in Europe."

7. Quoted in William Drozdiak, "Behind Show of Unity, Splits Threaten Allies," *International Herald Tribune* (Neuilly-sur-Seine, France), 27 March 1999.

8. Richard Holbrooke, *To End a War* (New York: Random House, 1998), 160.

9. Quoted in Ann Devroy and Bradley Graham, "NATO Suspends Bombing in Bosnia; Serbs Pledge to Withdraw Heavy Weapons, Agree to Talks Aimed at Cease-Fire," *The Washington Post,* 15 September 1995.

10. Madeleine Albright, "SECSTATE Testimony before HIRC Panel February 25," February 25, 1999 (http://www.eucom.mil/europe/serbia_and_montenegro/kosovo/usis/1999/february/99feb25a.htm), accessed 22 April 2001.

11. Quoted in "Chirac rejects Yugoslav ceasefire offer," *Agence France Presse,* 6 April 1999.

12. Quoted in "British minister says Kosovo brutalities must stop," *Agence France Presse,* 6 April 1999.

13. General Wesley Clark (Ret.), "Press Conference," 1 April 1999 (http://www.nato.int/kosovo/press/p990401c.htm), accessed 5 April 2001.

14. General Wesley Clark (Ret.), "Press Conference," 25 March 1999 (http://www.nato.int/kosovo/press/p990325a.htm), accessed 15 April 2001.

15. "NATO's Inner Kosovo Conflict," 20 August 1999, (http://news.bbc.co.uk/hi/english/world/europe/newsid_425000/425468.stm), accessed 8 July 2003.

16. Quoted in Hugh Dellios, "While Government Backs NATO, Greek Hearts Go with Serbs," *Chicago Tribune,* 16 April 1999.

17. "NATO Working on Ceasefire Proposal for Kosovo, Bonn Says," *Deutsche Presse-Agentur,* 31 March 1999.

18. Quoted in "Schroeder Rejects Any Change in NATO Strategy," *Turkish Daily News,* 20 May 1999.

19. In spite of the request, Novi Sad was subject to frequent attacks.

20. See William Drozdiak, "Air Strikes Jolt New Eastern Allies," *International Herald Tribune* (Neuilly-sur-Seine, France), 13 April 1999.

21. General Wesley K. Clark (Ret.), interview on security issues, *Policy.Com,* 8 August 2000.

22. Quoted in Darren Lake, "Kosovo Report Highlights NATO's Shortcomings," *Jane's Defence Weekly,* 1 November 2000, (http://www4.janes.com/kz/doc.jsp?kzdockey=/content1), accessed 14 July 2003.

23. Clark, interview, *Policy.com.*

24. George Robertson, "Kosovo One Year On," (http://www.nato.int/kosovo/repo2000/better.htm), accessed 8 July 2003.

25. Lieutenant General Michael C. Short, testimony, Senate Armed Services Committee, *Lessons Learned from Military Operations and Relief Efforts in Kosovo*, 106th Cong., 1st sess., 21 October 1999, 402.

26. Linda D. Kozaryn, "Cohen, Shelton Say NATO's Patience, Precision Paid Off," *American Forces Press Service*, (http://www.defenselink.mil/news/Jun1999/n06111999_9906113.html), accessed 8 July 2003.

27. Quoted in "Yugoslavian Air-Defence System Withdrawn from Kosovo," *Jane's Missiles and Rockets*, 1 July 1999, (http://janes.com), accessed 25 July 2003.

28. "Serbs Launched SAMs at Record Rates," *Jane's Missiles and Rockets*, 1 March 2000, (http://janes.com), accessed 25 July 2003.

29. William M. Arken and Bogdan Ivonisevic, "Civilian Deaths in the NATO Air Campaign," *Human Rights Watch*, February 2000, (http://www.hrw.org/reports/2000/nato/Natbm200.htm#P39_994), accessed 8 July 2003.

30. Robertson, "Kosovo One Year On."

31. *Lessons Learned from the Kosovo Conflict—The Effect of the Operation on Both Deployed/Non-Deployed Forces and on Future Modernization Plans: Hearing before the Military Procurement Subcommittee of the Committee on Armed Services*, House of Representatives, 106th Cong., 1st sess (http://commdocs.house.gov/committees/security/has292020.000/has292020_0f.htm).

32. Quoted in Ed Foster, "Imbalance of Power," *Jane's Defence Weekly*, 5 January 2000, 42.

33. "Serbs Launched SAMs at Record Rates."

34. "Yugoslavian Air-Defence System Withdrawn from Kosovo."

35. See Figure 16 of "Non-U.S. Aircraft Participating in Operation Allied Force" in Department of Defense, *Kosovo/Operation Allied Force after Action Report: Report to Congress*, 106th Cong., 2nd sess., 31 January 2000, 78.

36. See Figure 5 of "Allied Force Lessons Learned" in Department of Defense, *Kosovo/Operation Allied Force after Action Report: Report to Congress*, 106th Cong., 2nd sess., 31 January 2000.

37. Department of Defense, *Kosovo/Operation Allied Force after Action Report: Report to Congress*, 106th Cong., 2nd sess., 31 January 2000, xx.

38. Clark, interview, *Policy.com*.

39. Javier Solana, "Statement by NATO Secretary-General on Air Strikes," 23 March 1999 (http://abcnews.go.com/sections/world/dailynews/solanatranscript.html), accessed 10 November 2001.

40. Alan Little, "Behind the Kosovo Crisis," *BBC News*, 12 March 2000, (http://news.bbc.co.uk/hi/english/world/europe/newsid_674000/674056.stm), accessed 8 July 2003.

41. Rachel Riley, "The World Has Failed the Kosovar Refugees," *Wall Street Journal Europe*, 9 April 1999 (http://www.hrw.org/hrw/campaigns/kosovo98/

wsjoped.shtml), accessed 10 April 2001; U.S. Department of State, *Erasing History: Ethnic Cleansing in Kosovo* (http://www.reliefweb.int/hcic/links/links_main.htm), accessed 10 April 2001.

42. Ushtria Çlirimtare e Kosovës (UÇK) is known in English as the Kosovo Liberation Army (KLA). Formed in the mid-1990s, the UCK advocated a campaign of armed insurgency against the Serbian authorities. In mid-1996, the UCK launched attacks on Yugoslav and Serbian police forces.

43. Department of Defense, *Kosovo/Operation Allied Force after Action Report: Report to Congress*, 60.

44. International Court of Justice, "Case Concerning Legality of Use of Force (Yugoslavia v. Belgium); Request for the Indication of Provisional Measures," International Court of Justice Website, (http://www.icj-cij.org/icjwww/idecisions/htm), accessed 21 June 1999; reprint, Washington D.C.: International Legal Materials, July 1999.

45. Ibid.

46. "ICJ Rejects Yugoslavia's Request for Order to Halt Use of Force by Belgium, Remains Seized of Case," *M2 Presswire*, 4 June 1999.

47. See "Rambouillet Agreement," (http://www.state.gov/www/regions/eur/ksvo_rambouillet_text.html), accessed 8 July 2003.

48. Robertson, "Kosovo One Year On."

49. By occupying the Pristina airfield, Russia guaranteed itself a place in the planned peacekeeping force for Kosovo. In the early days after the bombing stopped, Russian troops in Pristina refused NATO access to the airfield. Despite Russian intransigence, NATO forces supplied the Russian troops with food and water while the negotiations occurred. Russia attempted to use the airfield as a bargaining chip to get its own zone; however, NATO refused because it did not want a de facto partitioning of Kosovo into a Serb sector protected by the Russian force and an Albanian sector protected by the NATO force. After a week of negotiations, NATO and Russia agreed that Russian participation in KFOR would be limited to 3,500 soldiers, Russian peacekeepers would not get their own zone but would operate within a zone led by a NATO member, and Russian forces would not be subordinate to NATO commanders but operate independently.

50. Quoted in BBC News, "Generals Clashed over Kosovo Raid," 2 August 1999 (http://news.bbc.co.uk/hi/english/world/europe/newsid_409000/409576.stm#top). Under NATO rules of command, subordinate officers can appeal to their own Minister of Defense before carrying out orders.

51. "The Alliance's Strategic Concept," *NATO Press Release*, 23 April 1999, (http://www.nato.int/docu/pr/1999/p99-065e.htm), accessed 8 July 2003.

52. Article V: The Parties agree that an armed attack against one or more of them in Europe or North America shall be considered an attack against them all and consequently they agree that, if such an armed attack occurs, each of them, in exercise of the right of individual or collective self-defence recognised by Article 51 of the Charter of the United Nations, will assist the Party or Parties so attacked

by taking forthwith, individually and in concert with the other Parties, such action as it deems necessary, including the use of armed force, to restore and maintain the security of the North Atlantic area. Article IV: The Parties will consult together whenever, in the opinion of any of them, the territorial integrity, political independence or security of any of the Parties is threatened. "The North Atlantic Treaty," 4 April 1949, *NATO*, (http://www.nato.int/docu/basics.htm), accessed 8 July 2003.

53. Quoted in "NATO, British Leaders Allege 'Genocide' in Kosovo," CNN. 29 March 1999 (http://www.cnn.com/WORLD/europe/9903/29/refugees.01/).

54. Quoted in "Blair Praises 'Committed' Clinton amid Talk of Rift," *Daily Mail,* 17 May 1999.

55. Department of Defense, *Kosovo/Operation Allied Force after Action Report: Report to Congress*, 85.

56. General Wesley K. Clark, "Effectiveness and Determination," *NATO*, 2 June 1999, (http://www.nato.int/kosovo/articles/a990602a.htm), accessed 8 July 2003.

Chapter 4

THE POLITICS OF AIR STRIKES

Scott A. Cooper

In their book *Thinking in Time: The Uses of History for Decision Makers,* Harvard professors Richard Neustadt and Ernest May make an important observation. Washington decision makers, and even academics, students, journalists, and the average citizen, "use history in their decisions, at least for advocacy or for comfort, whether they knew any or not."[1] While most of their work concentrates on the question of whether decision makers, within the limits of their circumstances, could have done better, it also focuses on how decision makers often misread cases in history and draw inaccurate comparisons and parallels. Munich framed many decisions after World War II. Vietnam has been the military's frame of reference for over two decades, and the past decade has seen the Gulf War used as the antithetical comparison to Vietnam. Whether these analogies are appropriate, they are used over and over, often to the detriment of thoughtful reflection. The military itself indulges too often in the complacency of hindsight, and it has done so again in looking back on the Kosovo air campaign.

Much of the debate since Operation Allied Force, especially in military circles and the Air Force in particular, has centered around the dissatisfaction of many commanders with the strategy of the campaign. Many commanders are critical of the basic strategy choices made by NATO's leaders, arguing that politicians needlessly hampered the application of a coherent and doctrinally pure air power strategy, thereby risking American credibility and also prolonging the war itself. What is most disturbing about this after-action chastisement is the absence of the appropriate collegiality coupled with civilian primacy that is necessary for both healthy civil-military relations and good national policy. Exacerbating this the military's misreading of both the Vietnam War and the Gulf War. Vietnam is remembered as a case of air power being undermined by civilian control of air operations, with images of President Johnson and Secretary of Defense McNamara on their knees in the Oval Office selecting targets. The Gulf War is remembered as a textbook case of proper civilian non-involvement, with President Bush, Secretary of Defense Cheney, and others merely standing back while the air planners conducted a lethal and successful strategy.

Both of these notions are incorrect, and they are especially harmful because they lead to the subsequent conclusion that politicians should only set objectives, not involve themselves with military plans or scrutinize the conduct of operations. Studying Vietnam, Iraq, and Kosovo more closely will help illuminate the historical role of the relationship between civilian policy makers and military leaders in air strategy. This will hopefully contribute to a much-needed dialogue between military leaders and their civilian masters that goes beyond the current simplistic notion that civilians should only set objectives and then stay out of the way. Candor, collegiality, and a common sense of purpose are the fundamentals to success.

THE CRITICISM

Lieutenant General Michael Short, now retired, who served as the Air Component Commander during Allied Force, has publicly decried the strategy of an incremental, gradual escalation, appealing to the President and those above him, including the Commanders in Chief (CINC's), that they should heed the advice of aviators, who best understand how to carry out a campaign. He declared that "as an airman, I'd have done this a whole lot differently than I was allowed to do. We could have done this differently. We should have done this differently."[2] He further expanded his argument in a speech at the Air Force Association Air Warfare Symposium:

> We need to prepare our politicians as best we can for what is going to happen. If we are going to initiate an air campaign, not an air effort, but an air campaign, airmen need to be given the chance to explain what is going to happen to our political leadership. Airmen, who have practiced their craft and their trade for 30 or 35 years, need to be given the opportunity to make

that explanation. I read in General Horner's [the air component commander in *Desert Storm*] superb book how he went to Camp David and briefed the President of the United States on how he intended to conduct an air campaign to prepare the battlefield in Kuwait and Iraq. I am not campaigning for a trip to Camp David, but there was a case to be made for an air campaign, and airmen should have made that case.[3]

When this does not occur, as he claims it did not in Operation Allied Force, we end up with random bombing of military targets that lose the intent of effects-based targeting. He has claimed that if he had been allowed to "go downtown" and bomb targets in Belgrade immediately, he could have shortened the war by four weeks. Moreover, he has concluded that civilian policy makers not only do not understand air power, but should not hamper operations once committed: "Our politicians need to understand that we will do our best to make air power clean and painless as they want us to, but it is not going to work out that way.... When they choose to employ us, to take us to war, when they choose to use military force to solve a problem that politicians could not, then they need to grit their teeth and stay with us."[4] He decries the ad hoc campaign that was executed like a pickup game:

> Our targeting philosophy clearly has to be agreed upon before we start.... We need to have agreed how we intend to employ our forces. I am not so naïve as to believe that we will be able to execute an air campaign just because our nation wants to. But we need to have made that case, and if that case is not accepted, we need to have a fallback plan that works and gets it done. Again, we don't want to do this by happenstance. We want to do it by design.[5]

Perhaps his harshest criticism has been an oft-repeated line to his superiors in the name of his pilots, "Sir, don't risk lives to demonstrate resolve."

General John Jumper, who commanded U.S. Air Forces in Europe during Allied Force and is now the Commander of Air Combat Command, has voiced similar criticisms. He calls the 1990s the "era of the limited objective," with military operations fraught with caution and half measures. He compared the operation order of General Eisenhower for Operation Overlord in World War II to the complicated and vague guidance of Kosovo. Eisenhower ordered his subordinates, "You will enter the continent of Europe and, in conjunction with the other United Nations, undertake operations aimed at the heart of Germany and the destruction of her armed forces."[6]

Admiral Leighton Smith, now retired and the former commander of NATO forces in the Balkans, declared that Allied Force was "possibly the worst way we employed our military forces in history."[7] During the air campaign, another unnamed general referred back to Operation Instant Thunder, the initial plan for the Gulf War, complaining, "This is not Instant Thunder, it's more like Constant Drizzle."[8] There was a feeling among many in the military that the erratic pace of the campaign, especially with the target approval process, was undermining its effectiveness.

This criticism is useful, but it should not be read as a case of civilian micromanagement or ignorance about the efficacy of air power. Such criticism instead should illuminate the many challenges of fighting as a coalition, of the changes technology has wrought, and of the unique circumstances surrounding the use of force in Kosovo.

TECHNOLOGY CHANGES THINGS

Technology has changed warfare in many ways, but among the most significant is the ability for all levels of authority to scrutinize and to involve themselves in the battle itself. This must be examined thoroughly when considering the conduct of an air campaign. If a pilot in Vietnam was given a target to attack, he flew the mission and debriefed his flight, not unlike a mission in Allied Force. The difference is that the debriefing during Vietnam would have been only the pilot's recollection of events. Today we have the ability to reconstruct what happened, often with precise detail.

For instance, during one mission in Allied Force the crew flying an F-15E Strike Eagle was given the target of a bridge near Nis in Serbia. The weapon was an AGM-130, a propelled 2000-pound bomb that is dropped more than twenty miles from the target and is guided via television datalink from the cockpit of the aircraft. The mission was a success; the bridge was destroyed. Tragically, there was a passenger train crossing the bridge when the bomb impacted, and the postflight video shows exactly that. As a result of that attack, as General Short testified, "the guidance for attacking bridges in the future was: You will no longer attack bridges in daylight, you will no longer attack bridges on weekends or market days or holidays. In fact, you will only attack bridges between 10 o'clock at night and 4 o'clock in the morning."[9]

This fact of technology must be taken into account by all actors at all levels of government. In a recent speech, General Jumper summarized the situation well: "Here we put this young man in this situation where he knows that this bomb is enroute to the target, and the videotape that is recording in the cockpit is running, that an hour after he leaves that tape is going to be graded by the Commander of the United States Air Forces in Europe, the Supreme Allied Commander Europe, and probably the President of the United States."[10] We cannot brush aside this dilemma with the simpleton claim, "Let the war fighters fight," because that is an unacceptable solution. But so are immediate reaction decisions like the one to which General Short alluded. Whether that decision was appropriate is debatable, but unless this technological reality is addressed, thoughtful and rational decisions about targeting might not be made.

VIETNAM

The view of Vietnam among many in the Air Force was summarized by Lieutenant General Short during an interview soon after the Kosovo campaign: "For years (in Vietnam) we bombed a little bit, and then we backed off, and...had pauses, and so on. Then finally we sent the B-52s north around January of 1973, and lo and behold, we brought them to the table."[11] This has led to a common belief among aviators that the United States might have won the war in Vietnam had they been allowed to run it.

The air strategy in Vietnam under President Lyndon Johnson, Operation Rolling Thunder, lasted from 1965 to 1968. It is often characterized as a failure undermined by divergent strategies advocated by the civilian and military leadership: the military continually advocating more bombing and the civilians pushing for a more restrained policy of gradualism. That assessment is incorrect and is thoroughly spelled out by Robert A. Pape, in his book *Bombing to Win: Air Power and Coercion in War.*[12]

Pape details the nearly four-year air campaign against North Vietnam under President Johnson in which three different and competing air strategies were tried at some point, each advocated by different constituencies in the administration or the military. The first strategy was one of coercing North Vietnam by threatening its population and economy through limited bombing of its industrial economy and population with gradually increasing risk. This was advocated by Defense Secretary McNamara, his assistant John McNaughton, Joint Chiefs of Staff Chairman Maxwell Taylor,

Director of Central Intelligence John McCone, Ambassador Henry Cabot Lodge, Deputy National Security Advisor Walt W. Rostow, and Assistant Secretary of State William Bundy. This strategy was executed throughout the spring and summer of 1965, where bombing focused on a list of fixed targets. General Taylor summed up this strategy as "a gradual, orchestrated acceleration of tempo measured in terms of frequency, size, number and/or geographic location.... An upward trend in any or all of these forms of intensity will convey signals which, in combination, should present to the DRV [Democratic Republic of Vietnam] leaders a vision of inevitable, ultimate destruction if they do not change their ways."[13]

The second strategy "aimed to exploit military vulnerabilities, thwarting Hanoi's ability to succeed on the battlefields of the South. The main supporter of this strategy was JCS Chairman General Earle G. Wheeler, who had replaced Taylor in August 1964. Army Chief of Staff General Harold K. Johnson also backed it, as did theater commanders such as General William Momyer and Admiral U.S. Grant Sharp."[14] This strategy was pursued from the summer 1965 through the winter of 1966–67, with strike aircraft tasked with air interdiction in an effort to disrupt the North's infiltration of men and supplies. Pilots were given complete freedom for armed reconnaissance and reattacks of previously struck targets throughout North Vietnam except for small areas around Hanoi, Haiphong, and the Chinese border.

The third strategy was one "which also focused on civilian vulnerabilities but aimed at raising current costs rather than future risks. The architect of this strategy was General Curtis E. LeMay, then Air Force Chief of Staff, assisted by his successor, General John P. McConnell. Battlefield commanders also lent their support to this strategy."[15] This strategy was executed from spring to fall 1967. President Johnson removed most of the political constraints on bombing. As one study pointed out, "the only remaining possibilities for increased military action against the North were mining and bombing of ports, bombing dikes and locks, and a land invasion of the North."[16]

It is indisputable that President Johnson and his advisors often directly supervised the selection of targets during the air campaign in Vietnam. It is a myth, however, that such detailed civilian control is inappropriate, that it was seen by the military as an ineffective method, or that it undermined the effectiveness of the air campaign. As Pape makes clear and other studies have shown, nearly all of North Vietnam's industrial war potential was destroyed. In fact, the air campaign in Vietnam "failed because none of the strategies could exert much leverage against North Vietnam."[17]

DESERT STORM

To many in the military, Operation Desert Storm was a textbook case of how to conduct a military operation. It had an easily defined objective: kicking Iraq out of Kuwait. Civilian decision makers gave broad political guidance, and from that guidance the military commanders were allowed to design a campaign without further meddling. In contrast to Vietnam, where American politicians directed the incremental and restricted use of force without clearly stated political objectives, in Iraq the commanders were allowed to use decisive and overwhelming force with few restrictions and for clear purposes.

This view is probably overstated, and it misrepresents both the strategic realities of Vietnam and the implausible confluence of circumstances surrounding the Gulf War. In Vietnam, there was always the fear of provoking a Soviet or Chinese intervention. No such threat existed in 1991. Although targets were not picked in the White House during Desert Storm, air planners were not given carte blanche to plan and to conduct the campaign.

In fact, two days before the beginning of the air campaign, Secretary of State James Baker and Undersecretary of State for Political Affairs Robert Kimmitt went over the target list with Secretary of Defense Dick Cheney and Chairman of the Joint Chiefs General Colin Powell. According to Kimmitt, "It was very clear to both Secretary Baker and me...that those political considerations that had been expressed, both at the Cabinet level and [in the NSC Deputies Committee], had been well taken into account, and we both left the meeting very comfortable from a political perspective."[18] Such a comment reflects more about the close working and mutually respectful relationship between civilian decision makers and the military than it does about the absence of meddling by civilians.

There were also instances of civilian involvement in the details of the air campaign, among them the Al Firdos bunker incident. It is a useful illustration of the restraints that will always be placed on the waging of war. In the first week of February, 1991, there was a network of potential command post bunkers that had not been targeted but that began to gain the attention of several intelligence analysts. Intelligence began to collect SIGINT—signals intelligence—that was emanating from the vicinity of the Al Firdos bunker in southwest Baghdad, and it was believed that it was being used by the Iraqi secret police. The bunker was put on the target list and was struck by two F-117s on the night of February 13. It is estimated that 204 civilians perished in the attack, all of whom had sought shelter in the Al Firdos bunker.

In the aftermath of the incident, General Powell and Rear Admiral Mike McConnell, intelligence director for the Joint Staff, went to the White House and defended the selection of the target to President Bush. Powell made it a policy thereafter to review all sorties proposed against the Iraqi capital. Commander in Chief of Central Command General Norman Schwarzkopf also required from then on that the air planners justify every mission in Baghdad beforehand, orally at first, and then in writing.[19] In the remaining two weeks of the war, only five targets were struck in Baghdad, all carefully chosen, as compared to twenty-five targets struck during the two previous weeks.[20]

Another circumstance of the war that illustrates the sometimes detailed civilian jurisdiction over air planning was the search for Scud missile launchers in Iraq. Only hours into the air campaign, Iraq launched several Scuds at Tel Aviv, Israel. American leaders from President Bush to Secretary Baker tried to convince the Israelis to show restraint and not to retaliate. Part of the argument they laid out for the Israelis was that there was nothing their air force could do that the American air force was not already doing. Even President Bush pledged a relentless American effort to destroy the Scud sites.[21] The order to suppress the Scuds was driven by Washington, not by the air planners in Riyadh. Air planners in Washington, even General Schwarzkopf himself, worried privately that the effort to destroy Scuds would hinder the main effort of the air campaign.

This was a contentious issue during the opening days of the air campaign, even leading to a heated exchange between General Powell and General Schwarzkopf, Schwarzkopf complaining of Washington meddling. But the guidance did not change. In fact, there was even a team of photo specialists sent to Israel to help interpret satellite images and recommend targets for American pilots.[22] The initial air campaign plan designated twenty-four F-15E Strike Eagles to suppress mobile Scud launchers. Eventually that number would triple, involving both F-16 Fighting Falcons and A-10 Warthogs. Daily detailed accounts of Scud-hunting activities were sent to Secretary Cheney at the Pentagon. There was even a plan to divert nearly all allied aircraft for three days of attacks against any site in western Iraq that could even remotely support Scud operations, although it was never carried out.[23]

The relative goodwill during Desert Storm between military air planners and the civilian masters is the result of the unique circumstances of the Gulf War, not that the civilians gave policy guidance and then butted out. First, there was a strong consensus among all countries involved, both at the military and political level, about the objectives of the war. There were almost no instances of cold feet among allies or political leaders in the United States. Second, there was little disagreement about how the air campaign should be carried out, unlike with Vietnam and Allied Force. Both the air planners and civilians in the White House were generally in agreement about the conduct of the campaign, with the mild exception of Scud hunting. Third, the environment of Iraq made for a much easier air campaign than any might have foreseen or than we can anticipate in the future. The jungles of Vietnam and the forests of Kosovo provided a much more difficult targeting problem than the deserts of Iraq, both from the perspective of the military effect of striking a target and regarding collateral damage considerations. Desert Storm was the easy case, and therefore it might not be the best example for future wars.

ALLIED FORCE

Any serious analysis of Allied Force must begin by considering the circumstances leading up to the decision to bomb Serbia. The agreement that was reached in October 1998 between Slobodan Milosevic and Richard Holbrooke was seen as a vehicle to buy time to reach a political settlement before the resumption of fighting that was expected in April. On January 15, 1999, Serb paramilitary and armed forces massacred at least forty-five people in Racak in southern Kosovo, blatantly violating the October agreement. This proved to be a turning point for the United States and NATO, although the NATO allies, with few exceptions, were not prepared to take military action. Finally, as military action appeared to be imminent after the failed talks at Rambouillet, there was widespread belief within the Clinton administration, among the NATO allies, and even within the military itself that decisive military action was not required. Most believed that a few days of bombing would coerce Milosevic to agree to a political deal. Even Secretary of State Madeleine Albright stated on the first night of the war, "I don't see this as a long-term operation."[24] The successful limited bombing in September 1996 in Bosnia that eventually led to the Dayton Peace Accords, Operation Deliberate Force, was what they expected would occur.

Unquestionably, there were also doubts both inside the military and the administration about the probability of successfully coercing Milosevic after only a few days of strikes. Shortly before the air campaign, the service chiefs testified before the Senate Armed Services Committee about their skepticism over whether air strikes by themselves would compel

Milosevic.[25] But also, all involved recognized that the imperatives of consensus politics, keeping all nineteen allies on board, ruled out a classic, decisive air campaign initially. As General Clark has admitted since the campaign, "no set of targets, and no bombing series was more important than maintaining the consensus of NATO."[26] When seen in this light, it becomes apparent that there was not a choice between using overwhelming force or not using force at all.

So the middle ground was tried initially, and had it worked, there would have been little discussion about the air strategy that was chosen. The first night of the campaign, March 24, 1999, saw 120 strike sorties attack forty Serbian targets. After a few days of the air campaign, it became apparent that Milosevic was not going to sue for peace, and General Clark received authorization from the North Atlantic Council to ramp up attacks against a broader spectrum of fixed targets in Serbia and fielded forces in Kosovo.

During the fourth week of the campaign, targeting efforts began to focus not just on the fielded forces in Kosovo but also on Milosevic's political machine, the media, the security forces, and the economic system, with such approved targets as national oil refineries, railway lines, road and rail bridges over the Danube, military communications sites, and factories capable of producing weapons and spare parts. By the end of the sixth week of the campaign, the bombing of infrastructure targets had cut Yugoslavia's economic output capability by half and had left more than 100,000 civilians out of jobs. Finally, during the final two weeks of the campaign, Serbia's electrical power-generating capacity was struck.

This escalation took place despite numerous obstacles: the reluctance of several alliance members; the process of sorting out procedures, authorities, and concepts of operations that had great effect on the target approval process; the lack of forces initially in theater; the hesitation of the U.S. administration and the Joint Staff to escalate; and finally the division among those in the U.S. military over the most appropriate targeting strategy. In fact, the disagreement within the military over strategy may have hampered the effectiveness of the air campaign more than any other factor. General Clark and Lieutenant General Short had a fundamental difference of opinion about the appropriate focus of the bombing. Clark believed the Serbian 3rd Army, the fielded forces in Kosovo, should be the focus of effort, while Short believed this was a waste of valuable munitions and sorties. Instead Short advocated bombing strategic targets that were valuable to Milosevic. The theory goes that these targets were the Achilles' heel of the enemy, that if destroyed the central leadership would be isolated and the enemy's military would collapse under light military pressure without guidance from above. In Allied Force, these targets were

Milosevic, his cronies, and the industries and buildings they personally valued, such as counterintelligence, security forces, loyal military units, and the related communications facilities. The result was a somewhat ad hoc campaign in the initial stages, with Clark's priorities generally prevailing, but one which eventually saw the expansion of all target sets throughout Kosovo and Serbia.

The air campaign also suffered several missteps that certainly hampered the achievement of an aggressive and uninterrupted strategy: the unfortunate bombing of a refugee column on April 14, the mistaken bombing of the Chinese embassy on May 7, and the tragic strike on a bridge over the Nis River when a passenger train was crossing the bridge. Despite all these, the air campaign proceeded and escalated rapidly.

This is not to say that it was the most effective strategy, but that the relationship between the alliance members, between the military and itself, and between the military and civilian policy makers proved a workable one over time. They ultimately found common ground, they stayed the course, and they prevailed. That a gradual, incremental strategy is not the best use of air power may not be as important as remembering that efficiency must sometimes be subordinated to political considerations. Those political considerations must be balanced against doctrine. That balancing act requires a close working relationship and frank dialogue between civilian policy makers and military professionals.

General Short has even admitted that the dialogue that took place about the conduct of the air campaign was a frank one. After the conflict, in his testimony before the Senate Armed Services Committee, he stated, "Certainly there are things that I believe could have been done differently, and I was given every opportunity to speak with my senior leadership about that. At no time was I prevented from expressing my thoughts."[27] And as General Clark stated after the campaign, "Once we crossed the threshold with the use of force, then my military colleagues and I had to speak up, and drive it toward the effective use of force."[28] The lessons of Kosovo are not to decry the incremental strategy or to bridle at political restrictions, but instead to recognize those limitations will exist and not allow ourselves to be fooled into believing it is "us" (the military) versus "them" (the politicians).

CONCLUSION

The end of the Cold War and the military operations of the 1990s reflect some important changes in the American way of war. Much of that change has to do with the fact that most military interventions of the past decade were not situations that immediately threatened the United States. It could even be argued that most were altruistically motivated, though certainly American interests were also involved. This reality has necessitated military restraint, one of the dilemmas of the modern era of warfare. Such restraint is understandable and makes us remember Clausewitz's classic words, "The political object is the goal, war is the means of reaching it, and the means can never be considered in isolation from their purposes."[29]

The notion that it is inappropriate for civilian leaders to involve themselves in the details of military operations is pervasive in the military. It is also misguided. Politically imposed rules of engagement and limitations on target selection will always be the servant of use of force decisions that conform to political objectives. Those political objectives are articulated by civilians. Strong civilian control of the military in a democracy is only self-limited, though civilian leaders should know when to show restraint. The candid and forthright civil-military relationship characterized by a shared sense of purpose is one of the linchpins of sound wartime policy.

NOTES

1. Richard E. Neustadt and Ernest R. May, *Thinking in Time: The Uses of History for Decision-Makers* (New York: The Free Press, 1986), xii.

2. William Drozdiak, "Allies Need Upgrade, General Says," *Washington Post,* 20 June 1999.

3. Lieutenant General Michael C. Short, USAF (Ret.), speech to AFA Air Warfare Symposium 2000, Orlando, Fla., 24 February 2000.

4. Ibid., 3.

5. Ibid., 7.

6. "Famous Quotations," (http://www.famous-quotations.com/asp/acquotes.asp?author=Dwight+D%2E+Eisenhower&category = All), accessed 20 December 2000.

7. "Reporters' Notebook," *Defense Week,* 19 July 1999, 4.

8. John D. Morrocco, David Fulghum, and Robert Wall, "Weather, Weapons Dearth Slow NATO Strikes," *Aviation Week and Space Technology,* 5 April 1999, 26.

9. Lieutenant General Michael C. Short, testimony, Senate Armed Services Committee, *To Receive Testimony on the Lessons Learned from the Military Operations Conducted As Part of Operation Allied Force, and Associated Relief Operations, with Respect to Kosovo,* 106th Cong., 1st sess., 21 October 1999, 402.

10. General John Jumper, "21st Century Aerospace Force: Essentials for Operational Success," speech to DFI International, Washington, D.C., 13 April 2000.

11. Lieutenant General Michael C. Short, interview, *Frontline,* Public Broadcasting System, 22 February 2000 (www.pbs.org/wgbh/pages/frontline/shows/kosovo/interviews/short.html).

12. Robert C. Pape, *Bombing to Win: Air Power and Coercion in War* (Ithaca, N.Y.: Cornell University Press, 1996), 174–95.

13. General Maxwell Taylor, quoted in *Pentagon Papers* 3: 316.

14. Pape, *Bombing to Win,* 181.

15. Pape, *Bombing to Win,* 180.

16. John E. Mueller, *War, Presidents, and Public Opinion* (New York: Wiley, 1973).

17. Pape, *Bombing to Win,* 189.

18. American Enterprise Institute, "The Gulf War Conference" (transcript, 7 December 1991), 236.

19. Rick Atkinson, *Crusade: The Untold Story of the Persian Gulf War* (Boston: Houghton Mifflin, 1993), 276–90.

20. Thomas A. Keaney and Eliot Cohen, *Revolution in Warfare? Air Power in the Persian Gulf* (Annapolis, Md.: Naval Institute Press, 1993), 185.

21. Atkinson, *Crusade,* 84–93.

22. Atkinson, *Crusade,* 131.

23. Atkinson, *Crusade,* 147.

24. Madeleine K. Albright, interview, *Newshour with Jim Lehrer,* Public Broadcasting System, 24 March 1999 (www.pbs.org/newshour/bb/europe/jan-june99/albright_3-24.html).

25. Bradley Graham, "Joint Chiefs Doubted Air Strategy," *Washington Post,* 5 April 1999.

26. General Wesley Clark, interview, *Frontline,* Public Broadcasting System, 22 February 2000 (www.pbs.org/wgbh/pages/frontline/shows/kosovo/interviews/clark.html).

27. Short, testimony, 398.

28. Clark, interview.

29. Carl von Clausewitz, *On War,* trans. and ed., Michael Howard and Peter Paret (Princeton, N.J.: Princeton University Press, 1984), 87.

Chapter 5

THE ETHICS OF PRECISION AIR POWER

Stephen D. Wrage

The joke among the B-17 crews in World War II was that they could hit any city, so long as it was large enough. Today, when one can strike a single building, or a floor of a building, or a window on that floor, it is hard to imagine air warfare of 1941 when whole cities were hard to locate and to hit. In light of today's standards—around 500 civilian deaths throughout the entire Kosovo campaign and no NATO air crew casualties in 38,000 sorties[1]—the manner of air war practiced sixty years ago seems immensely costly to the crews in the air and unthinkably brutal to their victims on the ground.

In fact, air warfare in 1941 was worse both for the flyers and for the noncombatants than can readily be grasped. Comparing then and now lets one focus on the extraordinary changes in the practice of air power that recent years have brought and highlights the exceptional growth that has occurred in both the public's expectations and in the policy-makers' temptations with regard to precision air power.

AIR WAR IN THE 1940S

A report prepared for Winston Churchill in August 1941 found that "of those aircraft attacking their targets, only one in three got within five miles...; over the Ruhr it was only one in ten."[2] That means that in the region where defenses were strongest, nine out of ten planes were failing to put their bombs within five miles of their targets, and the phrase "of those planes attacking their targets" should remind one of the many other planes that did not attack because they were damaged or downed by mechanical difficulties, ground fire, or enemy interceptors.

For almost a year both the British and the Germans had resisted targeting cities, but by winter of 1940 London and Munich, Coventry and Mannheim had been bombarded in city-for-city strikes purposely aimed at noncombatants. A grim score was kept in counts of civilians deliberately killed.[3] Unable to do better, they simply did their worst, aiming merely to produce corpses, to take the pain of war and the fear of death to the enemy and to give the home front a sense of blows being struck to make more bearable those being absorbed. The military gain in such ill-aimed acts of destruction was debatable at best, and the military cost was large. (Of the moral cost, more later.) According to John Keegan, British airmen died in such numbers in missions over France and Germany throughout 1941, and their bombs fell so randomly, that deaths among the crews in the air actually outnumbered the deaths they were able to inflict on their enemies on the ground. In short, the British airmen died in great numbers "largely to crater the German countryside."[4]

For a time the American B-17 crews, who joined the war in August 1942, did marginally better in terms of aiming at military objectives and away from noncombatants. They took the extra risk of carrying out daylight raids, and when they employed the best and newest techniques with Norden bombsights, they were able to do significant damage to targets such as shipyards and oil refineries, though they suffered severe losses to do so. But before long they reverted to the indiscriminant tactics of their British allies. Across a span of eighteen months, the bombers' penetration, accuracy, and survival rates all rose, and by the end of 1942, massive slaughter of noncombatants was being accomplished regularly. American and British air crews worked together in mid-1943 to produce 42,000 civilian deaths in Hamburg. That was 30 percent of the population of that city, incinerated in four nights of raids.

Churchill and Roosevelt seem to have felt little pressure to justify this mass killing of noncombatants. In early 1940, Churchill condemned "this odious form of warfare," but after Britain began purposely targeting civilians he dropped the condemnation and instead promised "the systematic shattering of German cities."[5] Roosevelt argued that burning cities would shorten the war, but even at the time that point was debated, and one finds few persons since then who will argue that line.[6] As one would expect in wartime, there were many not entirely coherent statements blaming the enemy ("Our plans are to bomb, burn and ruthlessly destroy in every way available to us the people responsible for creating the war"[7]) and many indistinct and figurative claims that the mass killing by incineration was exactly what ought to be happening—that the enemy should suffer "an ordeal the like of which has never been experienced by any country."[8] Statements of outright justification seem to have been rare, but announcements of intention simply took for granted the idea that war is war and the targeters had no choice. During the course of the war, the actions of the Bomber Command and the 8th Air Force brought heavy praise, in part because for the first three years of the war, the bombers were "the only instrument of force the Western powers had brought directly to bear against the territory of the Reich."[9]

The British and American leaders, public, and commentators seem simply to have silenced their moral sense and ignored the blatant fact that this sort of bombing "fails every aspect of the moral calculus."[10] Obliteration bombing seems to have been vocally opposed by very few, and the sole eloquent voice of protest this author has found is that of a Jesuit priest named John C. Ford who in 1944 starkly exposed the utter indefensibility of such mass slaughter of noncombatants.[11] A few months after he published his protest, U.S. bombers used 1,600 tons of improved jellied incendiaries (napalm) on the flammable stick and paper housing of Tokyo to produce fires that devoured 89,000 civilians in a single night (March 9, 1945).

There was some reflection and recoil at the end of the war. Occupying troops found the utter destruction of German and Japanese cities sobering to witness at ground level,[12] and the urge to attach blame to someone led to British Air Marshall "Bomber" Harris being the only officer of that rank *not* recognized with a peerage at war's end. For our purposes, what is remarkable is the huge gap between what was practiced then, and practiced to general applause, and what is acceptable now.

THEN AND NOW

In March 1945 celebration drowned out any expressions of misgivings at the news of the mass murder of almost 90,000 civilians in a single night,[13] while in 1999 in Kosovo air crews and targeting teams were held accountable for each unintended death. Then no hesitation was felt at burning entire cities, while more recently, in the Kosovo and Afghanistan campaigns, officers from the Judge Advocate General corps did "due diligence" on each proposed target to determine if it conformed to the laws of war and fell within boundaries set by all the NATO allies. At that time the public and the media took no particular interest in the damage our weapons did, expecting only that it should be as great as possible, whereas now every strike that lands off target is literally autopsied by inspection teams that take measurements, collect samples, and conduct a forensic analysis.[14]

This is not to argue that the people of 1945 were moral imbeciles whereas we today are enlightened beings. The deterioration in moral standards began in emergency and spiraled downward from there. In 1940, when the direct and intended targeting of civilians began, the British faced as an imminent and not unlikely event the prospect of being crushed by Nazi forces that had recently and swiftly overrun Belgium, Holland, and France and isolated 300,000 fleeing British forces at Dunkirk. Aerial bombardment occurred nightly, and invasion across the channel was a daily possibility. The British had every reason to expect that fascist domination would mean the end not just of their government but of their culture and even of the national community that was both the source and the expression of their values. This is precisely the situation Michael Walzer defines as a supreme emergency—the condition that permits the resort to any measure that serves to stave off the threatened destruction of the community. (Britain in 1940 through 1942 is in fact the very example Walzer offers of the rare condition of supreme emergency.[15]) The British literally had their backs to the wall and they faced true and dire necessity.

Moreover, at the time when the British breached the moral standard against targeting civilians and began to aim directly at cities, they lacked the technical ability to aim more narrowly. They were not capable, as the American air crews later sardonically admitted, of doing much better than hitting an entire city, and often (most often) did worse. Additionally, at that time and for a couple of years to follow, they lacked any other means of bringing the war to Germany.

That was the state of affairs in 1941, and it makes understandable and even justifiable the initial resort to attacks on cities. By 1943, however, when the British and the Americans were immolating German cities in massive raids that killed substantial portions of their populations, the state of supreme emergency was past, and the technical means to aim better were available to them. (Indeed, it was that improved capability to penetrate defenses and locate and hit the targets they chose—neighborhoods— that made those raids fatal to so many noncombatants.) Nonetheless, the mass incineration of civilians in Germany continued.

The Americans' behavior shows the damage that can follow the breaching of moral standards. The Americans followed the British lead in city immolation and followed it with a gusto that outdid their mentors. The standard against targeting civilians proved to be a kind of firebreak that, once breached, was totally overrun.

The Doolittle raid on Tokyo was not the product of supreme emergency, but neither was it one in a long train of attacks aimed explicitly against civilian neighborhoods. It was a one-time strike on the Japanese homeland aimed as well as was possible and intended not to establish a climate of terror but to make both the American and the Japanese peoples know Japan was not immune to war's destruction. In 1945, however, when mass murder was being inflicted with near impunity, the Americans showed no moral restraint. There was no supreme emergency, nor was there even good reason to believe those terror attacks would spare the United States having to invade the home islands, with all the deaths that would involve on both sides. General Sherman's argument—that the brutal methods he employed would bring the war to a more swift conclusion—does not plausibly apply. Without coherent or convincing strategic rationale, and with no discernible regard for the lives of noncombatants, they went ahead after the March 7 attack on Tokyo to incinerate sixty other cities in the same fashion by the end of July. In those raids they lost few planes to enemy action, but they put to death in horrid fashion hundreds of thousands of civilians.

In 1940, decision makers faced supreme emergency (initially) and breached all the standards; we today face no emergency and choose carefully where and how we will engage our forces. Then news coverage was reflexively favorable and little inclined to treat matters such as the morality of target selection. (Indeed, Walzer shows that most of the British populace was not even aware that cities and not military sites were being targeted![16]) Now the whole world is watching when America deploys its air power. Then they lacked the means (at the outset of city bombing) to do better, but technical change generated substantial new options for them,[17] even if in the heat of war they did not take up those options. We now possess the means to act with great discrimination and discretion, and our technological abilities are rapidly multiplying our options. More accurate, more flexible weapons are being developed, and precision weapons are making up larger and larger percentages of the weapons used.[18]

Obviously we can do much better now than they could then at upholding proper moral standards in the use of force, and in such matters "can" implies "ought." But just how much better *can* we do? How do the new options in the use of force presented by precision weaponry work in relation to the standards of moral behavior outlined in just war theory?[19]

The dramatic gun camera footage from the Gulf War excited expectations in many quarters, and these expectations provide the kind of incentive for policy makers to adhere to high standards that was missing almost sixty years ago when technological progress once before offered opportunities to perform to a higher standard. Those expectations may now prove hard to match, and the impressive capabilities unveiled in recent years have introduced temptations that may prove hard for policy makers to resist. This essay reviews what experience has shown about precision weapons and suggests tensions between the traditions of just war theory and the expectations they raise as well as the temptations they present.

PRECISION WEAPONS AND JUST WAR

The principal option that precision-guided munitions afford is they let the targeters choose exactly where he or she wants a weapon to go. This allows the user of force to be highly effective. It also encourages others to hold that person strictly accountable.

Accuracy matters in moral terms primarily because it allows one to aim narrowly at legitimate targets and carefully away from innocents and noncombatants. In this respect, the characteristics of precision weapons enable their users to conform to the standards for discrimination established in the just war tradition.

Greater accuracy also means that in addition to greater discrimination, greater effectiveness can be achieved. Moderate increases in accuracy yield large increases in effectiveness, and if (as has lately become possible) one can place the charge next to or on top of the target, one can employ a much smaller charge. This means that highly accurate weapons permit one to practice an economy in the use of force, using smaller warheads, carrying out fewer strikes, putting fewer people at risk, and cutting back the number of occasions for errors and so the likelihood of unintended damage or killing. This economy of force allows users of precision weapons to conform to the standards for proportionality established in the just war tradition. In short, greater accuracy means greater care can be taken and both of the *in bello* tests that are part of the just war tradition, that of discrimination and that of proportionality, can be more fully satisfied.[20]

THE ETHICS OF PRECISION AIR POWER

Greater accuracy has also given rise to greater expectations, greater temptations, and equivocal consequences. In general, the heightened expectations surrounding the aiming and delivery of these weapons has complicated and compromised their usability, and the novel temptations attached to them have created the potential for violations of traditional moral standards. To date, it is remarkable that policy makers appear not to have succumbed to these temptations, perhaps because the expectations and the visibility surrounding precision weapons are so high.

In the air campaign over Kosovo, for example, expectations of near perfection in aiming on the part of the American public, the international press, and the U.S. Congress left NATO Supreme Commander General Wesley Clark musing that the only arena in which he could lose the war in a single day was on the television screen. Fearing that major errors might lead the White House, Congress, or NATO allies to terminate Operation Allied Force, he attempted to produce an entirely error-free campaign—a standard of performance seldom required of commanders in past campaigns.[21]

In the first week of the air war only one civilian was killed. Nonetheless, the expectations were so high, the publicity so pervasive, and the sensitivity so great that by the fourth week of the campaign the notion was established that the air campaign was marked by error after error or, as the propagandists of the Milosevic regime described them, crime after crime.

Errors inevitably occurred, and "Clark had to devote a good part of his day to reassuring allied militaries and governments on issues of civilian collateral damage.... NATO was always on the defensive and never did succeed in putting Yugoslav's claims into perspective."[22] One of Clark's pilots mistook trucks filled with refugees near Djakovica on April 14, 1999, for a military convoy. Seventy-three civilians were killed by that single misdirected weapon, and "Belgrade...succeeded in conveying the impression that this was a regular occurrence."[23] Later another pilot released a missile at a bridge seconds before a train started across the span. "The pilots' identities were shielded from the press, so it was Clark himself who went out after the bridge bombing, and showed the press the horrible gun camera footage: how the weapon was released and a passenger train came into view, a split-second too late."[24] These errors led Clark to do what General Norman Schwarzkopf had done before him in Desert Storm after the strike on the Al Firdos bunker: to rule that all strikes in Belgrade must be personally approved by him.[25]

Paradoxically, then, these precision weapons are much more capable than conventional weapons of meeting very high moral standards in terms of discrimination, yet they are much harder to use than conventional weapons have been in past wars. They are found morally objectionable (and so politically risky) on precisely the grounds of their greatest strength, and they have these liabilities because of the revolution in expectations they have engendered. Even though the use of force in Kosovo and Afghanistan was almost certainly more discriminating than had ever been the case, and even though it was more limited in scope and so more proportionate than most past uses of force, it was more politically loaded and more subject to political termination.

Errors were intolerable in another sense. When a third great error was made in the Kosovo campaign—the strike on the Chinese embassy—the Chinese and others were certain that it was no mistake at all but rather a brutal case of message-sending by the Americans. Not one but two weapons struck the very floor in the building where the Chinese "journalists" were working, doing their intelligence gathering and analysis.[26] Clearly the Americans knew exactly what they were doing, the aggrieved parties and others concluded. They have let the unprecedented capabilities of these weapons go to their heads and surrendered to the temptation to use them for acts of intimidation and control.

The confident expectation that all strikes could be delivered precisely and unfailingly on target led NATO allies to realize they could deliver detailed "off-limits" lists to the NATO commander, thus complicating the use of these weapons still further. French President Jacques Chirac "boasted to a French reporter that if there were bridges still standing in Belgrade, it was thanks to him."[27]

NATO allies were not the only ones to enter the targeters' arena. There was a sad irony awaiting officers who resentfully recalled the Rolling Thunder air campaign over North Vietnam and the image of Robert McNamara and Lyndon Johnson poring over maps spread on the floor of the Oval Office. Such officers probably were pleased and relieved when, after Desert Storm, George Bush Sr. declared "By God, we've kicked the Vietnam syndrome once and for all," meaning that the rout of Iraqi forces had put an end both to unsatisfying and equivocal outcomes and to the sense that civilians' inputs hampered the military's efforts. President Bush was getting things precisely wrong, however. In fact, precision weapons—their accuracy and the procedures used to select and review targets for them—greatly multiply both the technical opportunities and the political incentives for civilian intervention.

Civilian inputs are likely to be inspired by two contradictory impulses: to demand that the military produce quick and decisive results, and to ensure that they generate no embarrassing incidents. In the Kosovo campaign there seems to have been more concern with avoiding embarrassments than with producing results, and so the pressures, on balance, favored care and restraint and so served the moral goals of discrimination and proportionality. In future campaigns against al Qaeda and Iraq, the balance may tip in the other direction.

Expectations were only half the story, however. Greater discrimination also brought temptations, largely resisted, for the users of precision weapons. It was tempting, for example to grow more ambitious in targeting. Targeters in the Iraq, Kosovo, and Afghanistan campaigns began at once to work to finer and finer tolerances in picking legitimate targets out of not-to-be-targeted areas. In Belgrade, missiles were used to knock down a communications tower located in a residential area. The target was attacked as though it were a peacetime demolition project: two legs of the tower were destroyed, tipping the tower in a direction that allowed it to fall near, but not on, a community of houses. The Serbian Interior Ministry, located in the same block as a hospital, was thoroughly destroyed without damage to the hospital.

Greater accuracy will not yield greater discrimination if it merely encourages more risk taking. In fact, however, the sensitivity to unintended deaths in Kosovo was so great that few or no occasions of excessive risk taking have been recorded. When these weapons failed it was due to misdirection because of poor intelligence or because of a fault in the functioning of the guidance system and not because targeters tried to split too fine a hair with them. The air power survey results are not yet conclusive in the case of Afghanistan, and different patterns may emerge from the war with Iraq.

Shaving tolerances closer and closer in the air war over Belgrade was necessary in part because the NATO forces ran out of targets early on. Without ground spotters, General Clark's forces lacked sufficient aim points, particularly ones of sufficient value to justify the use of an expensive cruise missile.[28] Moreover, his air commanders were powerfully tempted to use the accuracy they possessed to get more aggressive in their targeting, to "go downtown" as General Michael Short put it. Short had serious confrontations with Clark over the political restraints that Clark felt and Short's desire to be effective, even devastating.[29]

A second line of temptation (not entirely resisted) was the urge to use the weapons to assassinate persons such as Mullah Omar, the leader of the Taliban. In past decades assassination has carried a dark stigma, and most Americans probably imagine the practice is outlawed.[30] The practice is not illegal, but it certainly is politically loaded. Former presidential advisor George Stephanopoulos reports that "Of all the words you can't say in the modern White House...none is more taboo than 'assassination.' "[31] After September 11th, however, American popular sentiment raised no obstacle to using missiles to hunt and kill Osama Bin Laden and his closest associates.

Here the political pressure for results (and the military's own understandable desire to strike back) might undercut adherence to targeting principles. Evidence to date, however, indicates that the U.S. military leadership continued to be scrupulous in applying careful legal standards to qualify each target. In fact, there were complaints that the delays in obtaining a legal opinion on one occasion may have allowed Mullah Omar to escape a building shortly before it was struck.

One can argue that assassinating an individual, if one aims carefully at that individual and away from innocent persons, is vastly preferable in moral terms to making war on a country. The act meets the discrimination standard (provided one has overcome the immense practical difficulties in locating and isolating the target and one has accurate real-time intelligence to confirm that it is he, that he is where you think he is, and that he is alone) and it excels in the proportionality standard.[32]

One thinks of Operation Allied Force, a seventy-eight-day air campaign in which 38,000 sorties were flown in an attempt to coerce Slobodan Milosevic to stop the expulsion of Kosovars from their homes. In other words, force was applied carefully but heavily to all of Serbia for ten weeks in order to make one man change his mind. Would it not have been preferable simply to kill Milosevic?

The record shows that such efforts have not worked smoothly in the past. Strikes on Muammar Qaddafi in 1986 and on Mohammed Farah Aideed in 1993 did not succeed in eliminating either (though it has been argued that Qaddafi was deterred thereafter from sponsoring terrorist attacks). The manhunt for Manuel Noriega proved difficult even after the United States had invaded and occupied his country. Assassination attempts on Castro embarrassed the United States and reinforced his regime. The requirements of real-time intelligence are formidable, and the potential for error is great.

The experience of the government of Israel with its policy of targeted killings (often with missiles fired from helicopters) suggests that even with extraordinary intelligence capabilities one ends up making grievous errors and killing many innocents, particularly once the targets learn to avoid allowing themselves to be isolated. One suspects that if the U.S. attempts to imitate Israeli practice, it will find itself on a slippery slope to diminishing standards of discrimination.

In other words, the United States would find itself adopting tactics more suited to its terrorist adversaries. It is a well-known irony that one takes on the characteristics of an enemy in attempting to defeat him. There is probably more benefit in the long term in standing for values that include the repudiation of such tactics than in attempting to gain a reputation for being good at them. Besides, "tyrants and terrorists are likely to be better at this sort of thing than Americans" warns international lawyer Abraham Sofaer.[33] If assassination becomes American practice, Vice President Cheney and his successors may never be able to return from their "undisclosed locations," but perhaps that is already the case for the long term ahead. One part of our government, presumably the CIA, may always be crouched and waiting like a sniper, ready to take whatever shot the carelessness of our adversaries affords. They must be held to a standard where they take only exceptionally clear shots.

If, as may prove the case as technology progresses, weapons striking from satellites or unmanned aerial vehicles can be used to do the killing, then there may be a strong temptation to undertake assassinations with "plausible deniability." When the possibility of striking without being traced emerges, we must expect the shadowy persons carrying out the assassinations to abandon the scrupulous standards that have apparently governed their activities up to this point.

CONCLUSION

The practice of air power and the moral standards by which it is judged have come a long way from the practices and disregard of standards of World War II. Great changes in technology have greatly raised expectations, and obligations as well. In matters of ethics, "can" implies "ought." If one can be more discriminating, one ought to be so. Indeed one has a moral obligation to do so.

Clearly these weapons are far from perfect, yet perfectionism—the irrational expectation of near-perfect outcomes—has proven a problem distorting their proper use. The public ought not demand an impossible perfection of their users, and military and civilian leaders ought to be careful not to sell these weapons as capable of zero-error performance. Simply running the gun camera videos during the Desert Storm was very powerful selling.

It is appropriate and even helpful that groups such as Human Rights Watch hold the United States strictly accountable for the performance of each weapon fired in conflicts such as those in Kosovo and Afghanistan and in the attacks on al Qaeda. Such is the new reality of precision war. The technological advances that have changed popular expectations have at the same time begun to raise the standards that make up the doctrine of justice in war.

It was a lesson learned in Vietnam and again in the campaigns against Slobodan Milosevic, against the Taliban, against Saddam Hussein, and against al Qaeda that in a war for hearts and minds, justice matters. Within the many limitations discussed above, precision weapons do make it more possible to use force in accord with canons of justice.

NOTES

1. William Arkin, "Operation Allied Force, 'The Most Precise Application of Air Power in History' " in *War over Kosovo: Politics and Strategy in a Global Age,* eds. Andrew J. Bacevich and Eliot A. Cohen (New York: Columbia University Press, 2001), 21–22.

2. John Keegan, *The Second World War* (New York: Viking, 1989), 420.

3. For an account of the way the British and the Germans initiated the targeting of urban populations, see George Quester, "Strategic Bombing in the 1930s and 1940s," in *The Use of Force: International Politics and Foreign Policy,* eds. Robert Art and Kenneth Waltz, 2d ed. (Lanham, Md.: University Press of America, 1983).

4. Keegan, *The Second World War,* 420.

5. Churchill speaking before Commons in July 1943, quoted in John C. Ford, S.J., "The Morality of Obliteration Bombing," in *War and Morality,* ed. Richard A. Wasserstrom (Belmont, Calif.: Wadsworth, 1970), 139.

6. Keegan argues that a more discriminating campaign aiming at German production centers would have ended the war more quickly. Keegan, *The Second World War,* 415–35. Ward Thomas offers a great deal of evidence supporting that conclusion. Ward Thomas, *The Ethics of Destruction: Norms and Force in International Relations* (Ithaca, N.Y.: Cornell University Press, 2001), 87–146.

7. Brendan Bracken, Minister of Information, quoted in Ford, "The Morality of Obliteration Bombing," 139.

8. Churchill speaking before Commons on June 2, 1942, quoted in Ford, "The Morality of Obliteration Bombing," 139.

9. Keegan, *The Second World War,* 417.

10. The phrase is Bryan Hehir's from his essay "Kosovo: A War of Values and the Values of War," in *Kosovo: Contending Voices on Balkan Interventions,* ed. William Joseph Buckley (Cambridge, U.K.: Eerdmans, 2000), 400.

11. John C. Ford, S.J., "The Morality of Obliteration Bombing." This 1944 essay is reprinted in *War and Morality,* ed. Richard A. Wasserstrom (Belmont, Calif.: Wadsworth, 1970). Michael Walzer remarks, "It should be said that the campaign against terror bombing, run largely by pacifists, attracted very little popular support." Michael Walzer, *Just and Unjust Wars* (New York: Basic Books, 1977), 257.

12. For an account of the degree of devastation in France and Germany, see J. Glenn Gray, *The Warriors: Reflections on Men in Battle* (New York: Harper and Row, 1969).

13. The American public had long since become accepting of mass killing from the air. "Strange how quickly shocking things become casual things. There was horror aplenty over the early bombings of the war. Today we announce similar bombings...and we casually shrug it off." Editorial in *America* magazine, August, 1943. Quoted in Ward Thomas, *The Ethics of Destruction: Norms and Force in International Relations,* 88.

14. For a description of the kind of monitoring work done by such organizations as Human Rights Watch, see William M. Arkin, "Checking on Civilian Casualties," *The Washington Post,* 9 April 2002, A-1. Arkin, a retired intelligence officer, describes visiting bomb sites in Afghanistan, interviewing local persons, collecting evidence, and assembling comprehensive lists of munitions used, damage done, casualties caused, and circumstances left behind. Arkin works in informal collaboration with the U.S. military and attempts to reconstruct the events surrounding every strayed bomb.

15. See Walzer, *Just and Unjust Wars,* 251–63.

16. "[A]s late as 1944, according to other opinion surveys, the overwhelming majority of Britishers still believed that the raids were directed solely against military targets." Walzer, *Just and Unjust Wars,* 257.

17. See Keegan, *The Second World War,* 415, for a description of the dramatic improvements in the accuracy and effectiveness of aerial bombardment by January 1944 when British and American air attacks did great damage to the French railway system and so made the German army much less mobile and much less capable of fighting.

18. In the Gulf War approximately 7 percent of the air weapons used were precision guided. In the Kosovo air campaign the portion was around 35 percent. In Afghanistan the percentage was substantially higher still.

19. This essay employs just war theory merely as a device to order the discussion of precision weaponry. It does not apply just war theory to make judgments, nor does it assume that just war theory is flawless or above matters of interpretation. It adopts just war categories because they are part of a well developed, widely understood tradition of thought and therefore are useful as a framework for an argument.

20. Experience with precision weapons in the last twelve years confirms that force has been used with fewer unintended destructive consequences than in earlier instances. Human Rights Watch found that there were between 488 and 527

unintended deaths in Operation Allied Force over Kosovo. Human Rights Watch, *Civilian Deaths in the NATO Air Campaign,* (http://www.hrw.org/reports/2000/nato/), accessed 11 July 2003. In the far larger uses of force in the Gulf War the number of unintended deaths was around 3,000. Middle East Watch, *Needless Deaths in the Gulf War: Civilian Casualties during the Air Campaign and Violations of the Laws of War* (Washington: Human Rights Watch, 1991). See Ward Thomas, *The Ethics of Destruction: Norms and Force in International Relations,* 169, for evidence of dropping numbers of civilian deaths per ton of munitions since the introduction of precision guidance.

21. One comparable instance of perfectionist expectations might be the deployment of forces four years earlier to Haiti where the commander, Lieutenant General David C. Meade, feared a single casualty might endanger his mission. This constraint yielded a bizarre outcome: "The commanders of the 10th Mountain Division weren't letting any of their boys outside the gates...unless they traveled in a reinforced platoon—about forty guys, with mortars and armored vehicles—and were accompanied by either a battalion commander or an executive officer.... U.S. soldiers had invaded Haiti for the primary purpose of protecting themselves." Robert Shacochis, *The Immaculate Invasion* (New York: Viking, 1999), 254.

22. Arkin, "Operation Allied Force," 15.

23. Ibid.

24. Michael Ignatieff, *Virtual War: Kosovo and Beyond* (New York: Henry Holt, 2000), 101.

25. General Schwarzkopf allowed no strikes of any kind in Baghdad for five days after the error at the Al Firdos bunker. For the remainder of the war there were only five more strikes in Baghdad. Ward Thomas quotes the Gulf War Air Power Survey's conclusion that "To all intents and purposes the civilian losses ended the strategic air war campaign against targets in Baghdad," and this meant, Thomas says, the end of efforts to "decapitate" the Iraqi military. Ward Thomas, *The Ethics of Destruction: Norms and Force in International Relations,* 88. This conclusion is significant as reports of the plans for an upcoming second war with Iraq prominently feature such decapitating strikes.

26. James Kurth, "The First War of the Global Era," in *War over Kosovo: Politics and Strategy in a Global Age* (New York: Columbia University Press, 2001), 92.

27. Ignatieff, *Virtual War,* 104. These limitations appear to have had financial as well as humanitarian origins. Arkin reports "Telephone communications had been a major target in the Gulf War, but remained largely off-limits due to collateral-damage concerns and allied investments." Arkin, "Operation Allied Force," 24.

28. Ignatieff, *Virtual War,* 99.

29. For more on this tension, see the essay by Captain Scott Cooper in this volume titled "The Politics of Air Strikes." See also David Halberstam, *War in a Time of Peace: Bush, Clinton and the Generals* (New York: Scribner's, 2001), 445, for General Short's desire to (in his own words) "put out the lights in Belgrade by tar-

geting the military and civilian communications centers, the petroleum centers, and the transportation network."

30. The consensus of a variety of authors holds that assassination is not illegal under American law but rather that it has been ruled out under various executive orders issued by presidents since Gerald Ford. See Kenneth Pollack, *The Threatening Storm: The Case for Invading Iraq* (New York: Random House, 2002). See also Ward Thomas, *The Ethics of Destruction: Norms and Force in International Relations*, 49–50.

31. Stephanopoulos is quoted in Ward Thomas, *The Ethics of Destruction: Norms and Force in International Relations,* 49.

32. Assassination has always been limited by the problem of getting an assassin near the targeted person. Precision weapons ease that problem somewhat, but they complicate the problem of locating, identifying, and isolating the target.

33. Sofaer is quoted in Ward Thomas, *The Ethics of Destruction: Norms and Force in International Relations*, 84.

Chapter 6

CONCLUSION

Stephen D. Wrage

It remains to consider under what conditions, for what purposes, and in the context of what grand strategic assumptions precision-guided munitions are likely to be used in the near term. This concluding chapter offers a few observations on the prospects for immaculate war, then surveys the ways a number of thoughtful people have responded to the numerous, diverse, and complex potentialities of these weapons.

Precision-guided munitions obviously confer on the state that possesses them a great increase in power. Military historian Paul Kennedy was impressed enough to declare recently, "Nothing has ever existed like this disparity of power; nothing. One hears the distant rustle of military plans and feasibility studies by general staffs across the globe being torn up and dropped into the dustbin of history."[1]

Kennedy is probably right, provided he is speaking solely of those powers that employ general staffs to draw up military plans, that support large military establishments, and that produce feasibility studies. Opponents who are less well staffed and equipped and who do not have the high-value fixed assets that precision-guided munitions are most capable of destroying may be less intimidated, for the experience of the Iraq, Kosovo, and Afghanistan campaigns has shown that precision-guided munitions are most effective against fixed civilian infrastructure targets, less effective against opposing military forces, and least effective against irregular forces operating without fixed bases and home societies to be put at risk.[2]

Russia is one of those powers with general staffs and war plans. To the Russians, precision-guided munitions present a deeply troubling pair of features: they are quite useable and they are quite without effective conventional countermeasures. They are useable in the sense that one may use them without having to cross any firebreak such as the one between nuclear and conventional weapons. Indeed, they are more useable than less-accurate conventional weapons, since their unintended effects can be better controlled. They are without conventional countermeasures, meaning that one must attempt to deter rather than to defeat their use.

Brigadier General Charles J. Dunlap Jr., an insightful Air Force legal officer, writes:

> Russian generals fear that, in a general war, Western nations could employ such "smart munitions" to degrade Russian strategic forces, without ever having to go nuclear themselves. Consequently, said General Volkov, Russia "should enjoy the right to consider the first [enemy] use of precision weapons as the beginning of unrestricted nuclear war against it."[3]

America's adversaries might not be the only ones to find precision weapons unsettling in some of their characteristics. Their very usability might be a source of concern for some, and their very controllability might deepen the concern. General Henry Halleck, President Lincoln's chief of staff, who was both a jurist and a general, saw fit to praise a very different sort of weapon. We have a number of his letters to Dr. Francis Lieber, Lincoln's expert on the ethics and laws of war, and in one of these Halleck warmly praised the Minie ball.[4] This kind of bullet, which was fired from a rifle rather than a musket, was more accurate, carried farther, and produced larger, more frequently fatal wounds than the traditional round musket ball.

In contrast to today's precision weaponry, the introduction of the Minie ball dramatically increased casualties and magnified the horror of war. This, to General Halleck, was a virtue in the sense that it made resort to war a more extreme and dangerous choice, and it made the continuation of a war more costly and unpopular. Knowing battles would be terrible, one did not lightly embark on war. Finding war costly, one sought an end to it. By contrast, if leaders knew war would be as brief, as low cost, and as low risk as these precision air weapons promise to make it, they would resort to force too often and too recklessly, or so Halleck would say.[5]

Halleck would add that another reason one did not embark on war was that one never knew where war would lead. Clausewitz taught that the logic of war is to escape attempts to limit it, and the somber wisdom of Clausewitz (who had seen a good deal of war) is drilled into military people at academies such as the one that produced this book. For this reason, Halleck's and Clausewitz's anxious skepticism may lie deep behind the U.S. military's cautious response to calls to carry the campaign against terrorism forward by marching on Baghdad.

Nor is it outlandish to think that the war over Kosovo might have spun out of control. The Serbs had sympathetic and disgruntled supporters in their coreligionists, the Russians. Near the end of the war, "when a Russian tank column, attached to the NATO stabilization force in Bosnia, raced to Pristina airport in June 1999, and Wesley Clark, Supreme Commander, Europe, ordered General Michael Jackson, the British Commander of NATO troops in Kosovo, to prevent the Russians occupying the airport, Jackson refused."[6] Jackson later defended his insubordination to Clark saying, "I'm not going to start the Third World War for you."

Other voices focus on other potential drawbacks to these weapons. To a traditional realist such as Henry Kissinger, these weapons are in some respects quite unwelcome. A realist's guiding principle is that sovereignty ought to be respected, but these weapons transgress sovereignty with particular ease, stealth, and irresistibility. Aggression once meant armies crossing borders. Now an unseen cruise missile can profoundly violate sovereignty, and the bar to a decision maker contemplating such a violation is much lower than it was in the days when initiation of war required a major, highly visible mobilization. Think of how momentous a decision it was for Tsar Nicholas II to begin loading his forces on trains and moving them to the front in 1914, and compare that with the ease with which President Clinton was able to order a brief flurry of air attacks on Afghanistan in reprisal for the embassy bombings in Kenya and Tanzania. Kissinger would echo Clausewitz and warn that violations of sovereignty, especially acts of war, bring special risks. The missile attacks on Afghanistan were complete before anyone outside a small circle even knew they were under contemplation, and though it seemed that the matter then was closed, September 11th suggested that acts of war have a way of spawning consequences impossible to foresee.

The ability to designate oneself the hegemon and to go about setting the world aright is as enticing to some as it is frightening to others. Dr. Madeleine Albright, when she was U.S. Permanent Representative to the United Nations, was committed to restraining Slobodan Milosevic and pushed for military measures against him. General Colin Powell, then Chairman of the Joint Chiefs of Staff, counseled against. "What's the point of having this superb military that you're always talking about if we can't use it?" she asked, according to Powell's account in his memoirs. "I thought I would have an aneurysm," he reports.[7]

These remarkable weapons and the options they create undermine both the Weinberger Doctrine and the Powell Corollary regarding overwhelming force. The ease, speed, low risk, and high degree of detachment these weapons appear to allow permit policy makers to brush by such questions as "Does some vital national interest require us to fight?" "Do we need a major commitment of forces?" "Do we have a clearly defined and achievable objective?" "Are the Congress and the American people behind this move?" "Have all other means of dealing with this problem been exhausted?" With precision-guided munitions in the American arsenal, it was possible for Secretary of State Albright to prevail in her calls for two or three days of airstrikes to bring Milosevic back into line with the Dayton Accords, and it was those two or three days of strikes that grew unexpectedly into seventy-eight days of bombing and 38,000 sorties.

Persons of a far more conservative cast may be equally emboldened and excited by these weapons. Max Boot, editor of the *Wall Street Journal*'s editorial pages, calls on Americans to "adopt a bloody-minded attitude" and get on with the job of "imperial policing" world wide. "The Savage Wars of Peace," as he calls them in the title of his book,[8] will not be fought solely with precision weapons, but air power will lead the way and boots on the ground will follow.

A more moderate and cautious voice is that of Dr. Richard Haass. While at the Brookings Institution, before becoming director of policy planning at the State Department, Dr. Haass proposed that the United States ought to play the role of "The Reluctant Sheriff," not policing the world, exactly, but forming up posses to address limited situations and so maintain law and order.[9] Dr. Haass should be cautioned that precision weapons may make posses less possible. Colonel Philip S. Meilinger, formerly a dean at the Air War College, warns that even America's most militarily capable allies such as Britain do not have stealth aircraft, cannot provide the extraordinary degree of electronic counter measures U.S. pilots depend on, and do not have the Link-16 tactical data system that allows near-real-time tracking of mobile targets. Without these capabilities, true joint operations are impossible, and coordination with less technologically sophisticated allies is of course far more difficult and limited. As a result, almost all the lethal force applied from the air in Afghanistan was delivered by American planes and missiles, and a very unattractive division of

CONCLUSION

labor has been forced on those who would ally with the United States, where the United States provides "the air and space assets, which are seen as virtually impervious to enemy action, while U.S. allies supply the vulnerable ground troops."[10]

Dr. Charles Krauthammer thinks Haass is too timid. "It is time for us to seize the moment.... The United States astonished the [Arab] street with one of history's great shows of arms: destroying a regime 7,000 miles away, landlocked and far from American bases, solely with air power and a few soldiers on the ground.... The elementary truth that seems to elude the experts again and again—Gulf War, Afghan war, next war—is that power is its own reward. Victory changes everything, psychology above all. The psychology in the region is now one of fear and deep respect for American power. Now is the time to use it to deter, defeat or destroy the other regimes in the area that are host to radical Islamic terrorism."[11]

And who will stay behind to stabilize the region once the air strikes are over? Not the United States, says Krauthammer. "We don't peacekeep.... The United States should help the peacekeepers with logistical and, if necessary, air support. But *no peacekeeping troops.* [emphasis in the original]... [T]he American military is the world's premier fighting force, and ought to husband its resources for just that. Anybody can peacekeep; no one can do what we did in Afghanistan.... No to American peacekeepers. We fight the wars. Our friends should patrol the peace."[12] Presumably our allies will be as overawed as our adversaries and will accept the lackluster, hazardous, open-ended tasks of occupation and patrol. In Korea that task has meant staying at hair-trigger readiness for major war for almost fifty years.

Anatol Lieven of the Carnegie Endowment is made much less ebullient by these weapons. He contradicts Dr. Kennedy and recalls that the British enjoyed a similarly immense disparity of power over the Chinese in the Opium Wars. He describes two small frigates battering twenty-nine large war junks, sinking four, killing hundreds of Chinese, and losing not a man. Lieven offers two cautions in the spirit of hubris and nemesis: First, "the First Opium War so badly shook the prestige of the Manchu state that it became acutely vulnerable to attack by both bitterly anti-Western internal rebels and Britain's European rivals. Very soon, therefore, the British found themselves trying to prop up the regime that they had just defeated."[13] Second, he adds that the British, secure in their power, did not choose to reflect on the fact that they were defending a despicable and "indefensible trade, which China morally and legally had every right to ban."[14]

It is not hubris one should worry about but idealism, according to Dr. Alberto Coll. He sees in precision warfare a way for Americans across the political spectrum to indulge their penchant for Wilsonianism, which he sees as "exceptionalist, universalist, and messianic"—the conviction that "the United States is a unique society destined by Providence to lead the rest of the world to a future of freedom, democratic equality, and harmony."[15] To him, Albright, Boot, Haass, and Krauthammer are all Wilsonians, and Wilsonianism is an irresistible impulse at a time when the United States is "at the apex of its military, economic and political power"[16] and endowed with such useable devices as these for projecting that power. He would like to see more emphasis on simple "competence, foresight," and care not to "underestimate the adversary"—in short, a degree of "strategic humility."[17]

In any case, the vast disparity of power that Kennedy speaks of will not be permanent. "The technologies involved are neither abstruse nor expensive, and in time, America will lose its monopoly over them."[18] The American monopoly over nuclear technology, which was both more elaborate and more tightly held, did not last five years.[19] Long before rival countries can match Americans' technological feats with precision weapons, however, they will develop devices to neuter them. It cannot be too difficult to produce jamming devices to block or alter the weak signals broadcast by global positioning system (GPS) satellites. If these signals are blocked, joint direct attack munitions (JDAMs),[20] those most useful and cheapest of precision weapons, will stray from their targets and produce unacceptable unintended damage and deaths. Commanders will have no choice but to suspend their use.

Or perhaps rapid innovation, possibly simply in the strengthening of GPS signals, may keep precision weapons ahead of the efforts of those who would jam them. One measure of the rapidity of innovation is that the JDAM did not even exist in the American arsenal at the time of the Desert Storm campaign and was created out of dissatisfaction with the performance of laser-guided weapons in that dusty, frequently humid setting.

Perhaps the greatest barriers to other countries deploying precision warfare capabilities lie in the very high levels of skill and coordination required of the pilots, flight crews, intelligence officers, and all the other personnel involved in planning and carrying out air missions. These demands, along with the great costs involved, have been enough to bar all but a few countries from being able to deploy and operate air-

CONCLUSION

craft carriers. Perhaps they will prolong American dominance in precision air power.[21]

Over time, the potentialities of these weapons will be explored in large part by the military themselves. Since the earliest uses of guided munitions against bridges near Hanoi in 1975, but particularly since the Gulf War, American forces have been climbing a steep learning curve in the use of precision munitions and the related technologies. They have been innovating, experimenting with new methods, integrating new ways of doing things, and creating new tactics. There has been resistance to some missions, but those parts of the military with the most experience with precision air power have moved farthest away from a pattern of distrust and reluctance.[22]

After Desert Storm, it was not uncommon to hear "We do deserts; we don't do mountains," referring to a then-likely upcoming deployment to Yugoslavia. Many officers recalled that Tito's partisans had stopped divisions of the Wehrmacht and that the rough country and bad weather of the Slovene Alps offered many advantages to a defender. After pioneering innovation in the war over Kosovo, one tended to hear "Afghanistan will be harder." This was described as a graveyard for the Russians, and for the British long before them—for Alexander, too, for that matter. But there, too, pioneering adaptation, particularly coordination with ground forces, brought outstanding results. Before the second Iraq war, one heard the military caution that "We don't do urban warfare"—that although there have been many lessons learned since Mogadishu, the unavoidable realities of urban warfare tend to cancel out the advantages that precision weapons may bring.

The three-week military campaign against Iraq in 2003 proved that precision weapons can be devastatingly useful in urban warfare, particularly if traditional strategic air doctrine is set aside and air assets are instead employed as a kind of flying artillery, called in by spotters close to or above the scene. General Curtis LeMay would not have liked Operation Iraqi Freedom, but air power has never been better coordinated with troop movements or more effective in suppressing the enemy's fire and destroying his forces.

One still may hear, "We don't do countries with weapons of mass destruction." The words of an Indian general after Desert Storm are frequently recalled—"Don't fight the Americans without nuclear weapons."[23] As the search for Iraqi weapons of mass destruction continues, this dictum has not yet met its test. It is clear, however, that the preemptive strategy announced by the Bush administration in September 2002 was aimed at ensuring that those who possess such weapons do not have an opportunity to take the Indian general's advice. The talk of "shock and awe" attacks in Iraq turned out to be little more than talk, but the early penetration of Iraq by special operations forces working in conjunction with air power targeters clearly was intended in many cases to disrupt, intercept, and prevent the execution of Iraqi counterattacks. Whether or not these counterattacks could possibly have included weapons of mass destruction, the measures taken represent another significant innovation in the use of precision weapons and a major departure from the slow-build approach taken in Kosovo and the slow hundred-strike-per-day pace of the campaign over Afghanistan.

Perhaps these essays by the practitioners themselves will add a degree of realism, caution, and good sense to American thinking on waging precision war from the air. They are both the ones who actually apply the force—who endure the deployments, select the targets, spend the hours in the cockpits, take on the burdens and the risks of fulfilling the rules of engagement—and the ones who will have central contributions to make in devising the new ways in which missions will be accomplished. Their reservations and concerns expressed in this book might serve as some constraint on the readiness and eagerness of their leaders to set them to work applying force around the globe.

At present, it looks as though their commander in chief intends to use them often in pursuit of al Qaeda and to counter other countries that may find themselves placed on the "Axis of Evil." One hopes the president and his advisors will hear the cautions outlined in these pages by the practitioners who actually apply these weapons and perhaps maintain a healthy skepticism regarding the more sensational aspects of the promise of immaculate warfare.

NOTES

1. Quoted in Michael Kelly, "The American Way of War," *The Atlantic Monthly,* June 2002, 18.

2. See Darryl Press, "The Myth of Air Power in the Persian Gulf War and the Future of Warfare," *International Security* 26, no. 2 (2001): 5–44.

3. Charles J. Dunlap, Jr., *Technology and the 21st Century Battlefield: Recomplicating Moral Life for the Statesman and the Soldier* (Carlisle, Pa: Strategic Studies Institute, 1999), 6. Dunlap notes that in the Gulf War the Iraqis

responded to attack with precision weapons by setting oil fields afire, perhaps in hopes the smoke would blind the missiles and protect their troops and vehicles as they fled from Kuwait. Although no evidence has been found, as this book goes to print, that Saddam acquired weapons of mass destruction, America's precision weapons have given him, and other leaders who fear they may come under the American crosshairs, one more powerful incentive to try to achieve mass destructive capability.

4. Lieber produced the famous "General Orders One Hundred" on the law of armed conflict. General Halleck commissioned that study. Lieber and Halleck are discussed by James Turner Johnson, *Just War Tradition and the Restraint of War: A Moral and Historical Inquiry* (Princeton, N.J.: Princeton University Press, 1981), 289.

5. In fact, the three presidencies since precision guided munitions came into regular use in the Gulf War have all featured unusually high frequency of deployments of the military and applications of force.

6. Michael Ignatieff, *Virtual War: Kosovo and Beyond*, 207.

7. Colin L. Powell, *My American Journey* (New York: Random House, 1995), 576.

8. Max Boot, *The Savage Wars of Peace: Small Wars and the Rise of American Power* (New York: Basic Books, 2002).

9. Richard Haass, *The Reluctant Sheriff* (New York: Council on Foreign Relations Press, 1997).

10. Philip S. Meilinger, "Force Divider: How Military Technology Makes the United States Even More Unilateral," *Foreign Policy,* January-February 2002, 76–77. The degree of cooperation among allies in Operation Allied Force was less than has been generally understood. "According to Anthony Cordesman's careful study for the Center for Strategic and International Studies in Washington, the US flew 60 percent of all sorties, over 80 percent of strike sorties, over 90 percent of the electronic warfare missions and fired over 80 percent of the precision guided munitions and over 95 percent of the Cruise missiles. It is not simply that the Americans dominated the operational side of the mission; they also kept their NATO allies excluded from all targeting decisions involving American aircraft, and denied them intelligence for all targets struck by American missiles or planes," Ignatieff, 206. William Arkin confirms those figures and the practice of U.S.-only planning backed up by withheld intelligence. William Arkin, "Operation Allied Force, 'The Most Precise Application of Air Power in History,'" 3, 21.

11. Charles Krauthammer, "Victory Changes Everything," *The Washington Post,* 30 November 2001, A41. It is amusing to find in voices from the Left, particularly in Noam Chomsky and Edward Said, intimations that what Americans see as benevolent efforts to maintain order and stability in the world may be mere pretexts for the United States to use force to pursue selfish interests and have its way in the world. Krauthammer's blatant ambition to control the world scene makes their accusations seem badly understated.

12. Charles Krauthammer, "We Don't Peacekeep," *The Washington Post,* 18 December 2001, A27.

13. Anatol Lieven, "Hubris and Nemesis," in *War over Kosovo: Politics and Strategy in a Global Age*, eds. Andrew J. Bacevich and Eliot A. Cohen (New York: Columbia University Press, 2001), 99.

14. Lieven, "Hubris and Nemesis," 101.

15. Alberto Coll, "Kosovo and the Moral Burdens of Power," in *War over Kosovo: Politics and Strategy in a Global Age,* eds. Andrew J. Bacevich and Eliot A. Cohen (New York: Columbia University Press, 2001), 125.

16. Ibid., 124.

17. Ibid., 152, 153.

18. Ignatieff, *Virtual War: Kosovo and Beyond,* 210.

19. The United States tested nuclear weapons at Trinity in 1945. The first Soviet test was in 1949.

20. JDAMs are all-weather capable, inexpensive weapons bearing a ton of explosives. In Kosovo "[t]he $14,000 JDAM outperformed laser-guided bombs and cruise missiles that are 10 to 70 times more expensive, and became the weapon of choice for the most sensitive targets." Michael G. Vickers, "Revolution Deferred: Kosovo and the Transformation of War," in *War over Kosovo: Politics and Strategy in a Global Age,* eds. Andrew J. Bacevich and Eliot A. Cohen (New York: Columbia University Press, 2001), 194.

21. It should be noted, however, that some substantial and sophisticated elements of the targeting systems that support these weapons are already being duplicated by others. The *Economist* notes that Galileo, the European answer to the U.S. global positioning system, may be a step toward eroding America's exclusive dominance of precision weapons. "To Americans, [Galileo] smells of a European bid to end America's monopoly control of satellite-guided weaponry." "Europe's Galileo Satellite Positioning System: Eppur si muove—Or Maybe Not," *The London Economist,* 1 June 2002, 54.

22. Other parts of the military, especially the U.S. Army, have not. One instance of the Army's resistance to engagement in missions such as Operation Allied Force is the story of the long-delayed, ponderous deployment to Albania of Task Force Hawk, a contingent of twenty-four Apache helicopters. See Wesley Clark, *Waging Modern War,* 277–281, 303–5. The Army is not configured for rapid deployments in any case, but with Task Force Hawk they seemed to make the most of each obstacle as they slow-walked the aircraft toward the war zone. "To support and protect a mere 24 AH-64 Apache attack helicopters, the army determined that it was necessary to deploy a grand total of 6,200 troops. To provide this contingent with the wherewithal it required, the army shipped 26,000 tons of equipment to a staging area in Albania. Doing so consumed 550 C-17 sorties and cost $480 million. The cargo included more than a dozen 70-ton M1A1 tanks—too heavy to use on most Albanian roads—42 Bradley fighting vehicles, and 24 Multiple Launch Rocket Systems with extended-range, Army Tactical

Missile System missiles. To preside over this arsenal, the Army cobbled together a tactical headquarters that itself required the shipment of 20 five-ton Expando vans from Germany. The army also shipped 190 containers of ammunition, and enough repair kits to support twice the number of Apaches actually deployed. Thirty-seven other utility helicopters—Blackhawks and Chinooks—rounded out this mammoth task force." Michael G. Vickers, "Revolution Deferred," in *War over Kosovo: Politics and Strategy in a Global Age,* eds. Andrew J. Bacevich and Eliot A. Cohen (New York: Columbia University Press, 2001), 198.

23. The Indian general is quoted in Lawrence Freedman, *The Revolution in Strategic Affairs,* 45.

FURTHER READING

Atkinson, Rick. *Crusade: The Untold Story of the Persian Gulf War.* Boston: Houghton Mifflin, 1993.
Bacevich, Andrew J., and Eliot A. Cohen, eds. *War over Kosovo: Politics and Strategy in a Global Age.* New York: Columbia University Press, 2001.
Buckley, William Joseph. *Kosovo: Contending Voices on Balkan Interventions.* Grand Rapids, Mich.: William B. Eerdmans, 2000.
Byman, Daniel, et al. *Air Power as a Coercive Instrument.* Santa Monica, Calif., 1999.
Clark, Wesley K. *Waging Modern War: Bosnia, Kosovo and the Future of Combat.* New York: Public Affairs Press, 2001.
Dunlap, Charles J. Jr. *Technology and the 21st Century Battlefield: Recomplicating Moral Life for the Statesman and the Soldier.* Carlisle, Pa.: Strategic Studies Institute, 1999.
Fromkin, David. *Kosovo Crossing: American Ideals Meet Reality on the Balkan Battlefields.* New York: The Free Press, 1999.
Halberstam, David. *War in a Time of Peace: Bush, Clinton and the Generals.* New York: Scribner, 2001.
Holbrooke, Richard. *To End a War.* New York: Random House, 1998.
Ignatieff, Michael. *Virtual War: Kosovo and Beyond.* New York: Henry Holt and Company, 2000.
Kearney, Thomas A., and Eliot Cohen. *Revolution in Warfare?: Air Power in the Persian Gulf.* Annapolis, Md.: Naval Institute Press, 1993.
Luttwak, Edward. "Post-Heroic Armies." *Foreign Affairs,* July-August 1996.
Pape, Robert A. *Bombing to Win: Air Power and Coercion in War.* Ithaca, N.Y.: Cornell University Press, 1996.

Pollack, Kenneth. *Threatening Storm: The Case for Invading Iraq.* New York: Random House, 2002.

Press, Darryl. "The Myth of Air Power in the Persian Gulf War and the Future of Warfare." *International Security* 26, no. 2, (2001), 5–44.

Schelling, Thomas C. *Arms and Influence.* New Haven, Conn.: Yale University Press, 1966.

Thomas, Ward. *The Ethics of Destruction: Norms and Force in International Relations.* Ithaca, N.Y.: Cornell University Press, 2001.

INDEX

Aerospace warfare, 2
Afghanistan, 87, 93, 96, 101, 104
AGM-130, 74
Aideed, Mohammed Farah, 94
Al Firdos bunker, 77, 91, 98 n
Al Qaeda, 2, 96, 108
Albright, Secretary State Madeleine, 53, 79, 103, 104
Arkin, William, 96 n, 97 n, 109 n
Armed Services Committee, 57
Art, Robert J., 45 n, 96 n
Assassination, 94
Autopsies of bombed sites, 88
Axis of Evil, 108

B-52, 12
B-2, 57
Bacevich, Andrew J., 96 n, 110 n
Baker, Secretary of State James, 77
Balkans, 51–52
Belgium, 61
Belgrade, 61, 63, 65, 73
Bin Laden, Osama, 94
Boot, Max, 104, 109 n
Bosnia-Herzegovina, 52, 62, 64

Bowdoin, Gabrielle D., 48 n
Bracken, Brendan, 96 n
Brodie, Bernard, 24, 45 n
Buckley, William Joseph, 96 n
Bundy, McGeorge, 11
Bush, President George H. W., 72, 78, 92
Byman, Daniel, 31, 47 n, 49 n

Castro, Fidel, 94
Center for Strategic and International Studies (CSIS), 38
Centers of gravity, United States, 27
Cheney, Vice President Richard, 72, 77, 95
Chinese embassy in Belgrade, 13, 81, 92
Chirac, French President Jacques, 92
Churchill, Winston, 85, 86, 96 n
Cimbala, Stephen J., 46 n, 48 n
Circular error probable (CEP), 39, 49 n
Clark, General Wesley, 7, 12, 14–15, 17, 40–41, 49 n, 51–54, 56, 58, 63–64, 80, 91, 93, 103, 110 n
Clausewitz, Karl von, 82, 102, 103

Clinton Administration, 28
Clinton, President William, 7, 56, 60, 79, 103
Clodfelter, Mark, 48 n
Coercion, 23, 25, 45 n
Coercive diplomacy, 25
Cohen, Eliot A., 16, 96 n, 110 n
Cohen, Secretary of Defense William, 57, 64
Coll, Alberto, 106, 110 n
Collateral damage, 2
Collins, Joseph J., 48 n
Compellence, 25
Containment, 23
Cordesman, Anthony, 42, 49 n
CORDS (Civil Operations and Revolutionary Development Support), 17
Corley, Brigadier General John D. W., 5–6
Craig, Gordon A., 46 n
Crete, 54

D'Alema, Massimo, 55
Dayton Peace Accords, 10, 52, 79
Decapitation, 13
Decapitation, strategy of, 32
Denial, theory of, 31, 36
Deptula, Major General David A., 48n
Desert Storm, 92, 107
Designated mean point of impact (DMPI), 33, 41
Deterrence, 23, 25, 45 n
Dippold, Marc K., 47 n
Discrimination, 90, 94
Douhet, Giulio, 30, 47 n
Dunlap, General Charles J., 102, 108 n

Eisenhower, General Dwight, 73
Ellis, Admiral James, 41
Embassy bombings in Kenya and Tanzania, 103

Fischer, German Foreign Minister Joschke, 55

Fogleman, General Ronald, 10, 47 n
Ford, John C., S.J., 96 n, 97 n
Ford, President Gerald, 99 n
France, 59, 64
Freedman, Lawrence, 24, 26, 29, 45 n, 47 n, 111 n

George, Alexander, 24–26, 45 n, 46 n
Germany, 54–55, 59
Gorazde, 9
Gray, J. Glenn, 97 n
Greece, 54, 59

Haass, Richard, 49 n, 104
Halberstam, David, 98 n
Halleck, General Henry, 102, 109 n
Hallion, Richard P., 47 n, 48 n
Hehir, Bryan, 96 n
Hiroshima, 11
Holbrooke, Ambassador Richard, 10, 79
Horner, General Charles, 73
Human Rights Watch, 58, 96, 97 n
Hungary, 55, 59
Hussein, Saddam, 13, 96

Ignatieff, Michael, 98 n, 109 n, 110 n
Instant Thunder, 32, 36
International Court of Justice, 61
Iraq, 93, 101
Italy, 55, 59

Jackson, General Michael, 63, 103
Jakobsen, Peter Viggo, 26, 45 n, 46 n
JDAM, 57
Jervis, Robert, 46 n
Johnson, General Harold, 76
Johnson, James Turner, 109 n
Johnson, President Lyndon, 72, 75–76, 92
Joint Direct Attack Munitions, (JDAMs), 106, 110 n
Jumper, General John, 73, 75
Jus in bello, 11, 90

INDEX

Just war theory, 90, 97 n

Karadzic, Radovan, 9
Keegan, John, 86, 96
Kelly, Michael, 108 n
Kennedy, Paul, 101, 106
Kimmitt, Undersecretary of State Robert, 77
Kissinger, Henry, 103
Komer, Robert, 17
Kosovar Albanians, 53, 60, 62, 64–65
Kosovo, 52, 54, 57, 60, 63–65, 87, 91, 93, 101, 103
Krauthammer, Charles, 105, 109 n, 110 n
Kurth, James, 98 n

LeMay, General Curtis, 76, 107
Libya, 15
Lieber, Dr. Francis, 102, 109 n
Lieven, Anatol, 105, 110 n
Linebacker raids, 37
Link-16 tactical data system, 104
Lodge, Ambassador Henry Cabot, 76
Luttwak, Edward N., 47 n

Mahan, Alfred Thayer, 30, 47 n
Martonyi, Janos, 55
May, Ernest, 71
McCone, CIA Director John, 76
McConnell, General John, 76
McConnell, Rear Admiral Mike, 78
McNamara, Secretary of Defense Robert, 72, 75, 92
McNaughton, John, 75
Meade, General David C., 98 n
Meilinger, Philip, 3, 104, 109 n
Middle East Watch, 98 n
Milosevic, Slobodan, 2, 6, 21–22, 32, 35, 44, 51–53, 55–56, 62, 64–65, 79, 94, 96, 103, 104
Minie ball, 102
Mitchell, Brigadier General William "Billy," 31, 47 n

Mogadishu, 107
Momyer, General William, 76
Moral limits of coercion, 11–12
Mullah Omar, 93, 94

Nathan, James, 29, 47 n
NATO, 51–52, 54–59, 61–64
Neustadt, Richard, 71
Norden bombsight, 86
Noriega, Manuel, 13, 94
North Atlantic Council, 63

Obliteration bombing, 87
Operation Allied Force, 2, 21, 28, 42, 51, 58–59, 65, 71–73, 79–81, 91, 94, 98 n
Operation Deliberate Force, 9–11, 30, 79
Operation Desert Storm, 72, 77–79
Operation Enduring Freedom, 2, 3
Operation Horse Shoe, 7
Operation Instant Thunder, 74
Operation Iraqi Freedom, 3, 103, 104
Operation Overlord, 73
Operation Rolling Thunder, 75
Opium Wars, 105
Organization for Security and Cooperation in Europe, 52

Pape, Robert A., 30–33, 36, 44, 45 n, 46 n, 47 n, 48 n, 75–76
Parallel warfare, 32
Peacekeeping, 105
Pfaltzgraff, Robert L., 48 n
Pollack, Kenneth, 99 n
Powell Corollary to the Weinberger Doctrine, 16, 104
Powell, General Colin, 77–78, 103, 109 n
Precision guided munitions, 1, 90, 101
Press, Darryl, 108 n
Primakov, Yevgeny, 62
Pristina, 16, 62

Proportionality, 90
Punishment, theory of, 30, 36

Qadaffi, Muammar, 94
Quester, George, 96 n

Racak, Kosovo, 79
Rambouillet, 51, 61, 63
Ramstein Air Base, Germany, 5
Refugees, 7
Revolution in military affairs, 23
Riyadh, Saudi Arabia, 78
Robertson, George, 53, 56, 58, 62, 64
Rolling Thunder, 32, 37, 92
Roosevelt, President Franklin Delano, 86
Rostow, Deputy National Security Advisor Walt, 76
Russia, 52, 61–62, 66 n, 102

Sarajevo, 10
Schelling, Thomas, 8, 10, 24–25, 27, 31–32, 45 n, 46 n, 48 n
Schroeder, Gerhard, 55
Schultz, Kenneth, 47 n
Schwarzkopf, General Norman, 78, 91, 98 n
Sea control, theory of, 30
Serbia, 52, 54, 57, 59–61, 64
Sharp, Admiral U.S. Grant, 76
Shock and Awe attacks, 108
Short, Lieutenant General Michael, 12–15, 72, 74–75, 80–81, 93
Shultz, Richard H., 48 n
Simitris, Costas, 54
Singer, J. David, 46 n
Smith, Admiral Leighton, 74
Sofaer, Abraham, 95
Srebrenica, 9, 11
Stephanopoulos, George, 94, 99 n
Strategic coercion, theory of, 24
Supreme Allied Commander Europe, 52, 54, 64

Talbott, Strobe, 54, 58
Taliban, 96
Task Force Hawk, 110 n
Taylor, General Maxwell, 75
Technology, impact of on warfare, 74–75
Tel Aviv, Israel, 78
Thessaloníki, 54
Thinking in Time: The Uses of History for Decision Makers, 71
Thomas, Timothy L., 41, 49 n
Thomas, Ward, 96 n, 97 n, 98 n, 99 n
Tomahawk missile, 57
Trenchard, General Hugh, 31

United Kingdom, 65
U.S. Army Air Corps Tactical School (ACTS), 31
U.S. Naval Institute, 16
UNPROFOR, 52

Vickers, Michael G., 111 n
Vietnam Syndrome, 92
Vietnam War, 17, 32, 72, 75–77
Vojvodina, 55

Waltz, Kenneth, 96 n
Walzer, Michael, 11, 88, 97 n
Warden, Colonel John, 29, 32–33, 45, 48 n
Wasserstrom, Richard A., 96 n
Watts, Barry, 36, 48 n
Waxman, Matthew, 41, 47 n, 49 n
Weinberger Doctrine, 16, 47 n, 104
Wheeler, General Earle, 76
Wilsonianism, 106
World War II, 11, 17, 71
Wylie, J. C., 30, 47 n

Yamamoto, Admiral Isoroku, 13
Yugoslavia, 52, 55, 61–62, 65 n

Zinni, General Anthony, 16

ABOUT THE CONTRIBUTORS

SPENCER ABBOT is a U.S. Navy FA-18 pilot who has flown combat missions in Iraq and Afghanistan in conjunction with Operation Iraqi Freedom and Operation Enduring Freedom. Lieutenant Abbot is a graduate of the U.S. Naval Academy, where he served as Brigade Commander, and holds a masters degree in international relations from the Fletcher School of Law and Diplomacy. He is a Ph.D. candidate in international relations, and his dissertation examines the relationship between control and stability in the context of military intervention and post-conflict reconstruction.

SCOTT A. COOPER, a Major in the U.S. Marine Corps, recently returned from deployment to the Persian Gulf in support of Operation Iraqi Freedom. He wrote these articles as an International Affairs Fellow at the Council on Foreign Relations. He has served with the 22nd Marine Expeditionary Unit in support of Operation Enduring Freedom, and he has flown the EA-6B Prowler in support of the Iraqi no-fly zones, Operation Allied Force, and, most recently, Operation Iraqi Freedom. He is a graduate of the U.S. Naval Academy.

WILLIAM J. CROWE, in his more than 50 years of public service, was the United States ambassador to the Court of St. James', commander of all forces in the Pacific, chairman of the Joint Chiefs of Staff, and chairman of the president's Foreign Intelligence Advisory Board. Admiral Crowe earned a bachelor's degree from the Naval Academy, masters degrees from Princeton and Stanford, and a doctorate in politics from Princeton. Admiral Crowe currently serves as a distinguished professor at the U.S. Naval Academy.

DEREK S. REVERON received his doctorate from the University of Illinois at Chicago and a diploma from the Naval War College. During 1998–99, he served as a military and political analyst at NATO's Supreme Headquarters Allied Powers Europe (SHAPE) in Belgium where he distinguished himself during NATO actions in support of Kosovo. His first book, *Promoting Democracy in the Post-Soviet Region,* was published in 2002. In 2004, Palgrave Macmillan will publish *America's Viceroys: The Military and U.S. Foreign Policy.*

STEPHEN D. WRAGE is an associate professor in the political science department at the United States Naval Academy. He is a specialist in the formation of American foreign policy and writes on ethical issues in international affairs.

www.ingramcontent.com/pod-product-compliance
Lightning Source LLC
Chambersburg PA
CBHW051103230426
43667CB00013B/2424